Terra Aqua

This book is an anthology of key essays that foregrounds coasts, islands, and shorelines as central to the scholarship on the oceanic environment and climate across South Asia.

The volume is a collaborative effort amongst historians, anthropologists, and environmentalists to further understand the lifeworlds of the South Asian littoral that are neither fully aquatic or terrestrial, and inescapably both. *Terra Aqua* invokes a 'third surface' located in the interstice of land and water—deltas, estuaries, tidelands, beaches, swamps, sandbanks, and mudflats—and engages in a radical reconceptualization of coastal and shoreline terrains. The book explores uniquely endangered habitats and emergent templates of survival against rising seas and climatic disturbances with particular focus on the Bengal and Malabar coastlines.

A critical, transdisciplinary contribution to the study of climate change in South Asia, *Terra Aqua* examines salinity and submergence, coastal erosion, subterranean degradation, and the depletion of littoral lifeways impacting marine communities and biospheres. It will be of particular interest to scholars of environment studies, ecology and climate change in the Global South, hydrology, geography, ocean and island studies, environmental justice, colonialism, and imperial and maritime history.

Sudipta Sen is a professor of history and Middle East/South Asia studies, University of California, Davis. He is an author of *Empire of Free Trade: The English East India Company and the Making of the Colonial Marketplace* (1998); *Distant Sovereignty: National Imperialism and the Origins of British India* (2002); *Ganges: The Many Pasts of an Indian River* (2019) and a co-editor of the Routledge Ocean and Island Studies book series.

May Joseph is the founder of Harmattan Theatre and a professor of social science at Pratt Institute, and an author of the books *Ghosts of Lumumba* (2020); *Sealog: Indian Ocean to New York* (2019); *Fluid New York: Cosmopolitan Urbanism and the Green Imagination* (2013); and *Nomadic Identities: The Performance of Citizenship* (1999). Joseph is also co-editor of *Performing Hybridity* (1999). She co-edits three book series from Routledge: Critical Climate Studies, Ocean and Island Studies, and Kaleidoscope: Ethnography, Art, Architecture and Archaeology. Joseph creates site-specific performances along Dutch and Portuguese maritime routes exploring climate issues.

Ocean and Island Studies

Ocean and Island Studies is an interdisciplinary series concerning the role of oceans and islands in thought, theory, practice, and method, past and present. From remote island outposts to bustling island cities and the islands of our dreams, from the expanses and depths of the open sea to coasts, rivers, deltas, lakes, and polar icescapes, oceans and islands are at the core of much contemporary thinking. The books in the series explore the ways in which people use, envision, and construct marine, aquatic, littoral, island, and archipelagic geographies. It is a platform for Blue thinking from across the arts, humanities, social sciences, and environmental sciences. Meditations upon oceanic lyricism and altered states of 'islandness' find their place alongside research into the practicalities of island and coastal economies, infrastructures, and governance.

The series is open to book proposals from all segments of ocean studies, island studies, and related fields. It offers a mix of Shortform titles (i.e. Routledge Focus; 20,000-50,000 words) and Monographs (60,000-80,000 words), that is open to other text types as well.

Submissions contact: May Joseph, Managing Editor, routledgeoceanandislandstudies@gmail.com

Series editors:
May Joseph, Pratt Institute, USA
Adam Grydehøj, South China University of Technology, China
Philip Hayward, University of Technology Sydney, Australia
Sudipta Sen, University of California – Davis, USA
Lisa Bloom, University of California – Berkley, USA
Pamila Gupta, University of Witwatersrand, Johannesburg, South Africa

The Lives and Legacies of a Carceral Island
A Biographical History of Wadjemup/Rottnest Island
Ann Curthoys, Shino Konishi and Alexandra Ludewig

Terra Aqua
The Amphibious Lifeworlds of Coastal and Maritime South Asia
Edited by Sudipta Sen and May Joseph

For more information please visit: www.routledge.com/Ocean-and-Island-Studies/book-series/OISTU

Terra Aqua

The Amphibious Lifeworlds of
Coastal and Maritime South Asia

Edited by Sudipta Sen and May Joseph

LONDON AND NEW YORK

First published 2023
by Routledge
4 Park Square, Milton Park, Abingdon, Oxon OX14 4RN

and by Routledge
605 Third Avenue, New York, NY 10158

Routledge is an imprint of the Taylor & Francis Group, an informa business

British Library Cataloguing-in-Publication Data
A catalogue record for this book is available from the British Library

ISBN: 978-1-032-25276-6 (hbk)
ISBN: 978-1-032-25280-3 (pbk)
ISBN: 978-1-003-28247-1 (ebk)

DOI: 10.4324/9781003282471

Typeset in Times New Roman
by Newgen Publishing UK

To all dwellers of the terra aqua…

Contents

Figures

Contributors

Rohan D'Souza is a professor at the Graduate School of Asian and African Area Studies, Kyoto University. His PhD was awarded from the Centre for Historical Studies (Jawaharlal Nehru University). He was elected General Secretary of the Jawaharlal Nehru University Student's Union (1989–1990). He is the author of *Drowned and Dammed: Colonial Capitalism and Flood control in Eastern India* (2006) and his research interests include themes on environmental history and modern technology.

Naveeda Khan is an associate professor of anthropology at Johns Hopkins University. Her research spans riverine lives and national climate policy in Bangladesh, UN-led global climate governance processes, and Bengali and Urdu literature and writings on the environment. She is the author of *Muslim Becoming: Aspiration and Skepticism in Pakistan* (2012) and editor of *Beyond Crisis: Reevaluating Pakistan* (2010). Amongst her recent writings, a book manuscript "River Life and the Upspring of Nature" is in press and "Accounting for an Uncertain Future: the Paris Agreement and the Global South" is under contract.

Devika Shankar is an assistant professor of history at the University of Hong Kong. She is a historian of modern South Asia and the Indian Ocean region, and her research interests lie in the fields of environmental history, economic history, and science and technology studies. Her current book project focuses on the port of Cochin (now known as Kochi) on the south-west coast of India and examines how growing environmental concerns generated by the port's shifting coastline intersected with visions for its development in the late 19th and early 20th centuries. She has published articles on the history of water laws, princely sovereignty, and land acquisition in the South Asian context.

Preface

The plot for this collected volume of essays was hatched in 2020 during the height of the Covid-19 pandemic through a frenetic exchange of emails, phone calls, and impromptu zoom meetings. The concept of *Terra Aqua* emerged out of our shared concern with the future of the shorelines and coastal communities of Bengal and Malabar as visible sites of the ravages of climate change and sea rise. These conversations were further enriched by colleagues with whom we coedit the recently launched Routledge Ocean and Island Studies book series, dedicated to the notion that in the long shadow of the Anthropocene, as climate change and the weight of the planet's fossil-fuelled past alter sea levels, coastlines, ecosystems, and livelihoods, we need more than ever to try and fathom what changing oceans and oceanic spaces—shorelines, deltas, islands, and reefs—portend for the futures and pasts of the habitable world.

These essays are focused on the shorelines of the Arabian Sea and the Bay of Bengal, and on places and peoples that have long been left to the margins of history and ecology. The edge of the sea, as Rachel Carson reminded us, has always been an "elusive and indefinable boundary" endowed with a beguiling duality, belonging neither to land nor sea—a shifting world of tidal ebbs and floods that have produced over millennia a plethora of niches and habitats for plants and animals (Carson 1998). These enduring ecological formations, we argue in this volume, have also sustained foragers, hunter-gatherers, salt-makers, net-weavers, boaters, and fishers—generational custodians of the Terra Aqua. While *Terra Aqua* is inspired by the pressing immediacy of sinking coastlines and drowning ecologies of the Indian subcontinent, it is equally occupied with the long epochal pasts of tidelands, mangroves, sandbars, and mudflats.

The latest report of the Intergovernmental Panel on Climate Change's 2022 lays out the dire predictions of an uninhabitable future that is already underway, if the global community does not act immediately (IPCC 2022). The report brings to the fore the urgency of what Dipesh Chakrabarty calls "habitability". Much of the global South is changing into a space beyond the acceptable limits and understandings of habitability—as subsidence, submergence and swamplands take over the coastal regions of the world. Chakrabarty points out that "humans are not central to the problem of habitability, but habitability is central to human existence" (Chakrabarty 2021, 83). For much of the affluent global North, this habitability is being translated as infrastructure adaptation. For societies in the more populous and impoverished quarters of the Global South, this implies a perpetually insecure and unpredictable climate future. Caught in the interstitial space of the *Terra Aqua*—part mud, part saltwater—a large portion of the world's population in maritime Asia occupies this expanding zone beyond what Aldo Leopold once aptly described as the "land ethic", pointing out that land-use over the historical *longue durée* has demonstrated that the powerful largely determine what and who is valuable or worthless in planetary life (Leopold 1968, 204). Dwellers of the *Terra Aqua* are some of the first disappeared sites of such a dying ecumene.

Considering the most recent IPCC findings one can only assume that the global community has implicitly agreed that great swathes of the world's coastline and their amphibian inhabitants are simply expendable, resigned to the fact that such zones would soon slip into the realm of what Zygmunt Bauman calls "wasted lives" (Bauman 2004). This volume and its five case studies offer glimpses of such terraqueous spaces in South Asia. They remind us that these histories and ethnographies of abandonment are essential to the methods and praxes that must frame the misgivings of our collective planetary future.

References

Bauman, Zygmunt, *Wasted Lives: Modernity and Its Outcastes*. Oxford: Malden, MA: Polity; Distributed in the USA by Blackwell, 2004.

Carson, Rachel. *The Edge of the Sea*. Boston; New York: Houghton Mifflin, 1998.

Chakrabarty, Dipesh, *The Climate of History in a Planetary Age*. New York: Fordham University, 2021.

Intergovernmental Panel on Climate Change, 2022. https://report.ipcc.ch/ar6 wg2/pdf/IPCC_AR6_WG11_FinalDraft_Chapter10.pdf

Leopold, Aldo, *A Sand County Almanac and Sketches Here and There*. New York: Oxford University Press, 1968.

Acknowledgements

May Joseph would like to thank her mother Celine Netto for awakening her to the Kerala coast and teaching her how to live mindfully and sustainably by the sea. Thank you, Emily Briggs, and Lakshita Joshi of Routledge, for all the editorial support that went into making this book happen. To Sudipta Sen immeasurable thanks for keeping the mud and swamps of South Asia alive through the long years of Covid that generated this collaborative undertaking. Devika Shankar, Pamila Gupta, Naveeda Khan, Rohan D'Souza, thank you for joining the dialogue on South Asia's coasts. To Carl Zimring, Jennifer Telesca, Lisabeth During, Dan Boscov-Ellen, Francis Bradley, Todd Ayoung, Zhivka Valiavicharska, Luka Lucic, Martin Dege, Uzma Rizvi, Ira Livingston, Karin Shankar, Ariane Harrison, Jonas Coersmeier, Signe Nielson, Sameetah Agha, and Travis Holloway at Pratt Institute, thank you for your intellectual generosity in creating an environmentally engaged research world. Thank you to the Ocean and Island community of scholars, particularly Pamila Gupta, Adam Grydehoj, Lisa Bloom, Ilan Kelman, Godfrey Baldacchino, David Ludden, Smriti Srinivas, Neelima Jeychandran, Kavita Philip, Philip Hayward, Pedro Pombo, Arup Chatterjee, Gaurav Desai, Jonathan Pugh, David Shumway, Sofia Varino, El Glasberg, Charne Lavary, Dilip Menon, Brian McGrath, Edwige Tamalet, Brian Roberts, Mira Waits, Gwyneth Shanks, Brahim El Guabli, Jill Jarvis, Kristie Flannery, Renisa Mawani, Viju James, Jerome Whittington, and Jaimey Ferris, for opening up the oceans to me. Thank you to Ray Whritenour for sharing Lenape research on water and oceans. To Tavia Nyong'o, much appreciation for keeping the buried water sources of New York in sight as a guide to future thinking on where the water will go. Many thankyous to my local wine making, Chautu-Nadagam dancing, Luso-Malayali family along the Malabar coast whose generosity of spirit and resilience to hardships have inspired this inquiry. To Michael Taussig, R. Radhakrishnan, and

Alana Reuben, thank you for writing light. Special thanks to Gloria Zimmerman for keeping New York alive through Covid with walks along the Hudson River. Kate Neuman, always grateful for our ferry trips to the Rockaways that keeps the New York archipelago close. To my husband Geoff Rogers, gratitude for cultivating in me an appreciation for lichen, moss, stonewalls, and the flow of water. To my daughter Celine, immeasurable indebtedness for keeping me alert to why we must all continue to imagine a liveable world out of this terrible moment of climate insecurity.

Sudipta Sen would like to thank May Joseph for her steadfast support and collaborative zeal in putting together this volume during very challenging times. A big debt of gratitude to Devika Shankar, Pamila Gupta, Naveeda Khan, and Rohan D'Souza for contributions on and off the printed page. Thanks also to the Ocean and Island group of scholars including Adam Grydehoj, Lisa Bloom, Ilan Kelman, and Godfrey Baldacchino. Thanks to Debal Sen for his camerawork, and for making possible an unforgettable downriver exploration of the Sundarbans mangroves in the company of Indrajeet Samanta. Thanks to Nupur Dasgupta, Anuradha Ray, Samarpita Mitra and other colleagues at the Department of History, Jadavpur University, Kolkata where I explored the idea of an amphibious history of the greater Bengal delta in a series of invited lectures as a University Grants Commission, DSA senior visiting fellow in the September of 2019. And finally, a note of thanks to fellow-traveller, friend, and acclaimed author Parimal Bhattacharya, whose ideas and writings never cease to inspire.

Introduction

Sudipta Sen and May Joseph

We have been swept furiously into the era of the *Terra Aqua*, where the fusion of land and water is dramatically reshaping how we contemplate our planetary predicament. The temporal and spatial contours of the Anthropocene have brought into view with alarming clarity the pasts and futures of the coastlines and coastal waters of the world. New geographies and ecologies are manifesting where shores are disappearing into water, and coastline inhabitants are endlessly subject to the whims of the sea. As the planetary imprint of humanity transforms the behavior of oceans, chemically, biologically, and materially, it also reshapes and threatens in new ways the survival of shoreline communities across the world. Even beyond the pressing concerns of ecology and environment, the shifting equations between land and sea demand a reconsideration of the critical paradigms of both earth and water that are essential to human subsistence. *Terra Aqua* seeks to capture this critical juncture that encapsulates such life-transforming upheaval of the balance between the human and nonhuman forces that are reshaping the future of our planet. In many respects, it is the very threshold of the land and the sea where the most urgent challenges are unfolding for the people whose lives and livelihoods depend on the shorelines of oceanic South Asia.

This coedited anthology is a collaborative effort aimed at furthering the understanding, not only of lives at the edges of the landmass, but also *lives on edge* along what we describe as *Terra Aqua*—worlds that are neither fully aquatic or terrestrial, or that are inescapably both. Along with coastlines, a significant part of the world's island and archipelagic marine environments are located across the Global South, evident especially across maritime South Asia. We seek to address the precariat of these endangered habitats, reexamining the integuments of the natural, preternatural, and the biopolitical, and to question certain long-held, key assumptions about the morphogenesis and anthropogenesis of

DOI: 10.4324/9781003282471-1

coastal and island ecologies. Our task is to understand the new templates of survival against rising seas and climatic disturbances, salinity and submergence, coastal erosion, subterranean degradation, and the depletion of littoral lifeways across littoral communities and biospheres.

What are the most urgent stakes in thinking about sea-bound societies and terra-aquatic ecosystems and the communities that they sustain? How should we anticipate the coming shapes of environments, ecologies, bodies, livelihoods, exploitations, biohazards of the human and nonhuman worlds situated at the edge of the ocean and the mouths of deltas? What are the parameters of responsible, committed scholarship and ethical engagement in the study of coastal lives of the Anthropocene? If, as the science of global climate change suggests, the coastal and marine regions of South Asia are disproportionately affected by the warming of oceans, it is urgent that we recognize and disseminate voices from these vulnerable and marginal regions.

This collection features five scholars of riverine, oceanic, pelagic, and littoral ecologies and cultures of the Bay of Bengal and the Arabian Sea, who have undertaken to reconceptualize the idea of coastal terrains, along with the predicament of endangered littoral communities and cultures at the junctures of land and water beyond conventional geographical and ecological boundaries. They track how hinter-seas and inland littorals are foregrounding, changing global patterns of precipitation, melting, and flooding that have put distant seascapes of *Chiloe* of Chile and Beira in Mozambique on the same map of the Anthropocene. One of the most clearly visible signs of climate change and sea level rise is the dramatic shift taking place in the carbon-dioxide concentration, not only of the earth's atmosphere, but also oceans—evident in their rising acidity, which quickens the depletion of microorganismal precipitation of calcium carbonate, resulting in the catastrophic loss of coral reefs and coral ecosystems that once sustained a vibrant marine life, as they did millions of livelihoods—comprising in many cases the bulk of the national economy of islands and coastal countries. Since 1998, 90% of the coral has been lost in the Caribbean and the Indian Ocean, and the coral–algal tipping point today endangers, for instance, lives of fishermen from Belize to the Maldives. The extinction of coral reefs as safeguards against storm surges and flooding, endangering low-lying deltas and islands, epitomizes the fragile balance of habitats in the amphibian ecologies that we propose to bring into focus in this volume.

Littoral Frontiers

The oceans and seas of South Asia, much like those of East Asia, of the Mediterranean, and the Caribbean, are in rapid flux from acidified tides, melting ice, and monstrous storms. Such shifts in ocean ecologies and their foreseeable existential impact demand new categories of inquiry and analysis. Not only do they warrant an "oceanic turn" in environmental and climate change studies, but as Elizabeth DeLoughrey has urged, they also invite a new appreciation of the ocean as a disembodied and amorphous space, hard to reconcile within the familiar parameters of the lived human experience (DeLoughrey 2019). The idea of oceanic and terrestrial spaces as an unbroken continuum encapsulated in our theme "Terra Aqua" derives from this interstice of wet and dry habitations. The term "terraqueous" has been advanced recently by scholars such as Liam and Colás who argue that while global, transoceanic capital has redrawn coastlines and transformed marine ecosystems, the "liquid vastness" of the planet has never been permanently subjugated (Liam and Colás 2021, 3). Brian Roberts has also invoked the term terraqueous in a similar vein, noting the mutually constitutive limits of land and water in the making of geographical borders, and the sea as a continuum of the geopolitical terrain (Roberts 2021). Our concept of the *Terra Aqua* advanced in this book follows a different logic. While we share the idea of the ocean as a boundless, geodesical, political, philosophical, navigational, and epistemological space, we focus on the liminal spaces and intimate margins of inundated land, which are now at the center of the unfolding drama of sea level rise and its staggering environmental impact.

The upheaval of waterfronts affecting the coastlines of Pakistan, India, Bangladesh, Myanmar, and Sri Lanka urges a different awareness of the South Asian littoral as an extended, interconnected, and uniquely vulnerable ecological frontier. Such a reconfiguration of southern Asia, represented in this volume by the Malabar and Bengal coasts, helps us move beyond the naturalized geographical taxonomy of oceans and continents (Lewis and Wigen 1997). While in recent years the Indian Ocean world as a crucible of historical and cultural traversal and networks has emerged as a much-needed field of study, the vast edges of this subcontinental land mass have not received similar scrutiny. In repositing an alternative geo-historical imaginary for South Asia, we follow Greg Dening's suggestive description of the "wet stretch between land and sea"—not just as a threshold crossed by seafarers and inlanders, but indeed a world unto itself—fashioned out of sand, mud, swamp, and muck (Dening 2002, 9).

The history and ecology of the South Asian littoral also offer a valuable perspective on some of the largest climate-driven migrations taking place across the Global South from the Arakan coast to the Indonesian archipelago. Formerly stable coastal and island economies are facing the dire impacts of destroyed aquifers, sinking barriers islands, adding to the horrors of forced migration, most starkly evident in the case of the displaced Rohingya refugees of Myanmar (Uddin 2020). The contributors of this anthology, accordingly, explore the contours of archipelagos, lagoons, minor seas, hinter-seas, deltas, and estuaries that suggest new approaches to the shifting ecological landscape of coastlands dissolving and sinking in an increasingly wetter and hotter planet. The chapters in this volume center around the idea of earth and water as inseparable elements in the making of the human and nonhuman lifeworlds. It frames our understanding of subaltern protagonists in the Anthropocene—the putative bodies and subjects of the new *damnés* of the *Terra Aqua*. We also acknowledge here our intellectual debt to the political mobilizations that have redefined the Global South as an embattled shoreline, where the detritus of global, fossil-fueled, late industrial capital along with the new ravages of climate has subjected coastal peoples to a double marginality, socioeconomic and geographical.

Terra Aqua as a collaborative venture is an opportunity for this engaged group of scholars to explore the intersections of race, caste, gender, and reproduction at the oceanic margins inhabited by the traditional environmental subaltern—the poor, lower-caste, Dalit communities of South Asia, who all too often recede into the background as faceless victims in narratives of climate crisis. Dalit and other indigene fishing, boating, and hunting-gathering communities occupy and struggle to survive in some of the most inundated and waterlogged land along the South Asian and Indian Ocean littoral, and in every sense, they absorb the gravest risks of the anthropogenic and climatological changes affecting coasts and coastal waters.

Amphibious Terrains

Terra Aqua seeks to contribute to the new and burgeoning scholarship on water, waterbodies, riparian effluvia, sludge, and slime, through a critical review of the connective geohistorical matrix of pluvial, riparian, and littoral cultures shaped and reshaped by rivers, rains, winds, and ocean currents. Recent studies of the significance of rain and rainfall in human civilization (Barnett 2016) and the forces exerted by the dearth and excess of water on the monsoon-driven ecology of greater South Asia (Amrith 2018) are urging new directions in the study of water and wetness. Water from this viewpoint is thus more than a sociocultural artifact measured in terms of human need and value, but more significantly, a dynamic and creative entity that mediates all subsistence, especially at the margins of land and water. *Terra Aqua* takes a similar approach to the other constituent elements that define shoreline and backwater terrains, including salt, silt, sand, rock, plankton, mangrove, fish, and microbes. More importantly, it focuses on the interplay of land and water as a defining character of life and survival, expanding on what Gagné and Rasmussen have outlined as the amphibious anthropology peculiar to our times, responsive to the unprecedented transformation of terrains and habitations that pose serious challenges to conventional settings of ethnography and history (Gagné and Rasmussen 2016).

The chapters in this volume also take up the challenge of reconceptualizing the human geography of coastlines, following geographers Lahiri-Dutt and Samanta's study of life unfolding on the floodplain *chars* of India and Bangladesh (Lahiri-Dutt and Samanta 2013), and environmentalists Kraus and Harris (2021) who suggest that deltas need to be reconsidered as zones where "organic life and inorganic matter meet". We propose to extend this reappraisal of coasts and shorelines as an in-between, involuted world unto itself, and as a mélange of elements fresh and brackish, fluid and stagnant, human, and nonhuman—introducing a *third surface* of inquiry with land and water in balance. This analytic takes up Jason Cons' (2017) astute observation that we live indeed in an extended moment of global flooding, within the Anthropocene's biopolitical paradigm of periodic inundation that continually turns coastlines into unfamiliar and extended backwater spaces. Such a paradigm urges a reconsideration of new "hydrosocial" realms of human–environmental interaction (Kraus and Strang 2016; Linton and Budds 2014). Not just the in-between spaces of land and water, but also temporal rhythms attuned to aridity, wetness, and fluidity, and the dynamic interweaving of ideation and practice that define coastlines as living entities.

This idea of an interspersed realm of earth and water builds on our heightened awareness of the dynamic geomorphology and history of littoral spaces and habitations. It takes a cue from the maritime historian Michael Pearson who suggested that coastal edges of the world facing the open seas have much in common, offering a new paradigm for the study of littoral societies and their transoceanic connections (Pearson 2006). Pearson wrote about narrow strips of the Indian Ocean littoral as microcosms of lives and livelihoods adaptive to the perpetuum mobile of dunes, rocks, and flats exposed and submerged, and the constant advance and retreat of the land and the sea—ambiguous, shifting, and defying geographical definitions and cultural boundaries (Pearson 2003, 36–37). A comparable notion of the volatile ecology of foreshores has been advanced for our times by Meg Samuelson, who sees the coast as a constantly shifting frontier where "elements of earth and water ceaselessly overlap and draw apart", tied to diurnal and seasonal planetary motions (Samuelson 2017, 17). Our postulate of a third surface beyond the binary of land and water embraces the overlapping ecological and historical timescales of this in-between, saturated terrain.

Finally, *Terra Aqua* seeks to capture the changing granularity of these interzonal maritime spaces—a case in point being the vast delta islands of the South Asian littoral where the fluvial freshwater discharges collide with the saline currents of the Bay of Bengal, and where millennial processes of sedimentation and tidal erosion have created the largest halophytic mangrove forest in the world. This mangrove forest sits atop vast and ephemeral pelagic formations of mud and sand. These are strata and substrata neither fully liquid nor solid, "anti-pattern" (Bremner 2015), undifferentiated in their murky viscosity—unique backwater holdouts to the intensely cultivated and populated mainland.

The Five Chapters

The chapters in this book are dedicated to the exploration of amphibious terrains from three distinctive vantage points. First, they identify an interstitial, watery universe sculpted by silt and sand endowed with unique human geographies, where lives and livelihoods occupy liminal spaces, neither liquid nor firm, alternating between hunting-gathering and cultivation. They address how human settlements have continually reshaped such domains with plough and fishing net, boats, and paddle. Second, they recast coasts, shorelines, deltas, estuaries, sandbanks, and islands as unstable and dynamic ecological formations, rendered even more volatile from the point of view of survivors in the epoch of global warming, climate change, and rising sea levels. Third, they advance a deep ecology of this topography through a study of the subaltern bodies, whose struggles and ingenuity have for centuries defined and reconstituted these ever-changing margins of land and water.

The first two chapters in this collection are focused on the Arabian Sea and the west coast of India. May Joseph's chapter focuses on the Kerala floods of August 2018, that were attributed to excessive monsoon rainfall, which devastated Kerala's hinterlands and coastal regions, resulting in a terrible loss of lives and property. The deluge, aggravated by the opening of 35 dams without adequate warning to the residents, resulted not only in landslides and extensive soil erosion, but also in new levels of submergence of the Kerala coastline, making the pressing reality of Kerala's fragile and precarious shoreline along the Malabar Coast painfully evident. Joseph shows how the region between Kochi and Kollam became the space of the terra-aqueous, blurring the waterline between mud, sand, coastline, and dwelling place.

Parts of the urban habitations in this region became archipelago-like with their own flooded ecologies. Roadways were washed away, and the sea claimed stretches of the shore, with increasing levels of toxicity recorded in both water and sand. The topic of Devika Shankar's chapter also concerns the Malabar Coast. Her chapter explores the seasonal formations of the mudbanks that have always defined the profile of this volatile shoreline, including the geography and settlement of its port and harbor. Shankar writes about the interplay of water and sediment in the making and unmaking of this muddy and turgid terrain. She dexterously draws the geomorphology of the Malabar, inserting it into a new conversation with the human landscape of the coast.

The third chapter by Rohan D'Souza examines the legacy of colonial engineers in the service of the British Empire, whose descriptions of humid, tropical, monsoon-drenched, coastlines of provinces such as Bengal, alongside their statistical compilations and reports, laid out a land-based historical paradigm. Here, engineers, hydrographers, and surveyors sought to track the volatile geomorphology of river basins and the coastline from the vantage point of the *terra firma* rather than accepting the estuarine and tidal formations as an amphibian realm of land and water in perpetual flux.

The fourth and fifth chapters in the volume are set in the coastal areas of the Bay of Bengal and the extended deltaic, estuarine, and mangrove-rich Bengal shoreline that is now partof both India and Bangladesh. Naveeda Khan's chapter delves into the drowning of children in Bangladesh, which happens frequently along the sandbanks and temporary islands or *chars* that form in the middle of the Jamuna River, largest of the three major rivers in Bangladesh. Women attribute the death of their children to the work of supernatural beings such as Ganga Devi and Khwaja Khijir, reigning spirits of the water who instill forgetfulness and negligence in mothers and lure their children to the river's watery depths. Khan explores how such explanations of accidental death among the surviving poor of the riparian waterscape of Bangladesh offer a glimpse into the human dimension of climate change. The chapter by Sudipta Sen, which also studies the settlers of the aqueous terrain of the drainage basins of the Brahmaputra and the Ganga debouching into the Bay of Bengal, charts the deep ecology of the delta as a shifting and volatile geo-historical formation, where fossil remains of ancient peat-beds, freshwater and marine sediments, and carbonaceous clay form the matrix where people at the fringes of settled society have battled and survived natural elements—water, mud, and salt for centuries. Traditional settlers of this land such as the *Kaibartta*s and *Bagdi*s are some of the largest coastal caste groups of India, who still appear in the census with their subcaste status tied to their work as fisherfolk and guardians of levees and dykes. Sen's chapter is foray into the histories and lifeworld of people that colonial anthropologists once called the Dravidian "boat castes" and the "aboriginal" tribes of coastal Bengal as amphibious, subaltern figures, who are now the new precariat of climate change and ocean rise.

Lifeworlds of the *Terra Aqua*

As early as 1950, with characteristic prescience, Rachel Carson wrote about the oceans as our planetary thermostat—an "enveloping mantle" that radiates and absorbs heat, regulating and stabilizing the earth's temperatures through winds and ocean currents (Carson 1951, 172). In our current predicament of untenable temperatures and swelling sea tides, the marginal spaces between land and water and their long-term habitations on the brink of submergence are emerging as the new frontiers of survival. What affordances can we imagine for the unfolding realities of this hotter, wetter, marshier world? The chapters in this volume illustrate that life in the *Terra Aqua* of South Asia continues to demonstrate how human ingenuity has doggedly countered the harsh lessons of history. But how far will such struggles endure? We have crossed the point where the terrifying realities of the Bengal delta and the Malabar coast are mere symptoms of a planetary reality from which there is no place to hide. They are mirrors of our shared futures. Toward that end, these chapters are calls for a historical sense of a shared precarity and the unrelenting urgency of climate mitigation.

References

Abulafia, David. 2011 *The Great Sea: A Human History of the Mediterranean*. New York: Oxford University Press.

Abulafia, David. 2019 *The Human History of the Oceans*. London: Penguin.

Aikau, Hōkūlani, and Vicuna Gonzalez. 2019 "Curating a Decolonial Guide: The Detours Project." *SHIMA*, 13(2): 11–21. https://doi.org/10.21463/shima.13.2.04

Amrith, Sunil. 2018 *Unruly Waters: How Rains, Rivers, Coasts and Seas Have Shaped Asia's History*. New York: Basic Books.

Armitage, David, Alison Bashford, and Sujit Sivasundaram. Eds. 2018 *Oceanic Histories*. Cambridge: Cambridge University Press.

Arunachalam, B. Ed. 1998 *Essays in Maritime Studies*. Mumbai: Maritime History Society.

Baldacchino, Godfrey. 2004 "The Coming of Age of Island Studies." *Tijdschrift Voor Economische en Sociale Geografie*, 95(3): 272–283. https://doi.org/10.1111/j.1467-9663.2004.00307.x

Baldacchino, Godfrey. 2008 "Studying Islands: On Whose Terms? Some Epistemological and Methodological Challenges to the Pursuit of Island Studies." *Island Studies Journal* 3(1): 37–56.

Ball, Philip. 2014 *H2O: A Biography of Water*. London: Phoenix.

Ball, Philip. 2017 *The Water Kingdom: A Secret History of China*. Chicago: The University of Chicago Press.

Ballestero, Andrea. 2019 *A Future History of Water*. Durham: Duke University Press.

Barnett, Cynthia. 2016 *Rain: A Natural and Cultural History*. New York: Broadway Books.

Bauman, Zygmunt. 2004 *Wasted Lives: Modernity and Its Outcasts*. London: Polity.

Bishara, Fahad Ahmad. 2017 *A Sea of Debt: Law and Economic Life in the Western Indian Ocean, 1970–1950*. Cambridge: Cambridge University Press.

Bloom, Lisa. 2022 *Climate Change and the New Polar Aesthetics: Artists Reimagine the Arctic and Antarctic*. Durham: Duke University Press.

Boon, Sonja. 2019 *What the Oceans Remember: Searching for Belonging and Home*. Waterloo: Wilfred Laurier University.

Bremner, Lindsay. 2015 "Muddy Logics". In *Bracket 3: At Extremes*, edited by Lola Sheppard and Maya Przybylski. Barcelona: Actar: 199–206.

Campling, Liam, and Alejandro Colás. 2021 *Capitalism and the Sea: The Maritime Factor in the Making of the Modern World*. London, New York: Verso.

Carson, Rachel. 1941 *Under the Sea-Wind*. London: Penguin.

Carson, Rachel. 1951 *The Sea Around Us*. New York: Oxford University Press.

Carson, Rachel. 1998 *The Edge of the Sea*. New York: Houghton Mifflin Harcourt.

Chakrabarty, Dipesh. 2021 *The Climate of History in a Planetary Age*. Chicago: The University of Chicago Press.

Cons, Jason. 2017 "Global Flooding." *Anthropology Now* 9(3): 47–52.

Damodaran, Pradeep. 2014 *The Mullaperiyar Water War: The Dam that Divided Two States*. New Delhi: Rupa Publications.Davis, Mike. 2009 "Living on the Ice Shelf: Humanity's Meltdown." *The Nation Institute*. Posted June 26, 2008, www.tomdispatch.com/post/74949, accessed May 22, 2009.

DeLoughrey, Elizabeth M. 2007 *Routes and Roots: Navigating Caribbean and Pacific Island Literatures*. Honolulu: University of Hawaii Press.

DeLoughrey, Elizabeth M. 2019 *Allegories of the Anthropocene*. Durham: Duke University Press.

Dening, Greg. 2002 "Performing on the Beaches of the Mind: An Essay." *History and Theory* 41(1): 1–24.

De Villiers, Marq. 2000 *The Fate of Our Most Precious Resource*. New York: Houghton Mifflin.

D'Souza, Rohan. 2006 *Drowned and Dammed: Colonial Capitalism and Flood Control in Eastern India*. New Delhi: Oxford University Press.

Federici, Silvia. 2019 *Re-enchanting the World: Feminism and the Politics of the Commons*. New York: Autonomedia.

Finneran, Niall., and Christina Welch. 2020 "Mourning Balliceaux: Towards a Biography of a Caribbean Island of Death, Grief and Memory." *Island Studies Journal* 15(2): 255–272. https://doi.org/10.24043/isj.121

Gagné, Karine, and Mattias Borg Rasmussen. 2016 "Introduction – An Amphibious Anthropology: The Production of Place at the Confluence of Land and Water." *Anthropologica* 58(2): 135–149.

Gissen, David. 2009 *Subnature: Architecture's Other Environments*. Princeton: Princeton Architectural Press.

Gomez-Barris, Macarena, and May Joseph. 2019 "Coloniality and Islands." *SHIMA* 13(2): 1–10.

Goodell, Jeff. 2017 *The Water Will Come: Rising Seas, Sinking Cities, and the Remaking of the Civilized World*. New York: Little, Brown and Company.

Gooley, Tristan. 2016 *How to Read Water: Clues & Patterns from Puddles to the Sea*. London: Hachette.

Grydehøj, Adam, Xavier Barcelo Pinya, Gordon Cooke, Naciye Doratli, Ahmed Elewa, Ilan Kelman, Jonathan Pugh, Lea Schick, and R. Swaminathan. 2015 "Returning from the Horizon: Introducing Urban Island Studies." *Urban Island Studies* 1: 1–19.

Guerin, Ayasha. 2019 "Underground and At Sea: Black Marine Entanglements in the New York Archipelago". In *Coloniality and Islands*, edited by Macarena Gomez-Barris and May Joseph. *SHIMA* 13(2): 30–55.

Guha, Ramachandra. 2014 *Environmentalism: A Global History*. Gurgaon: Penguin India.

Gupta, Pamila. 2012 "Monsoon Fever." *Social Dynamics* 38(3): 516–527

Gupta, Pamila. 2021 "Ways of Seeing Wetness." *Wasafiri* 36(2): 37–47.

Hadjimichael, Maria, Costos M. Constantinou, and Marinos Papaioakeim, 2020 "Imagining Ro: On the Social life of Islets and the Politics of Islandography." *Island Studies Journal* 15(2): 219–236.

Haraway, Donna. 2015 "Anthropocene, Capitalocene, Plantationocene, Chthulucene: Making Kin." *Environmental Humanities* 6(1): 159–165.

Hau'ofa, Epeli. 1994 "Our Sea of Islands." *Contemporary Pacific* 6(1): 148–161.

Hong, Gang. 2020 "Islands of Memory, Islands of Trauma: The Case of Dongzhou, Hengyang, China." *Island Studies Journal*, 15(2): 237–254. https://doi.org/10.24043/isj.131

Horden, Peregrine, and Nicholas Purcell. 2000 *The Corrupting Sea: A Study of Mediterranean History*. London: Blackwell.

IPCC. 2022 Climate Change 2022: Impacts, Adaptation and Vulnerability (IPCC Sixth Assessment Report).

Iyer, Ramaswamy R. 2003 *Water: Perspectives, Issues, Concerns*. New Delhi: Sage.

Iyer, Ramaswamy R. 2007 *Towards Water Wisdom: Limits, Justice, Harmony*. New Delhi: Sage.

Joseph, May. 2019 *Sea Log: Indian Ocean to New York*. London: Routledge.

Kaiser, Birgit M., and Kathrin Thiele. 2017 "What Is Species Memory? Or, Humanism, Memory, and the Afterlives of '1492'." *Parallax* 23(4): 403–415. https://doi.org/10.1080/13534645.2017.1374510

King, Tiffany Lethabo. 2019 "Off littorality (Shoal 1.0): Black Study Off the Shores of 'the Black Body'." *Propter Nos* 3 (winter): 40–50.

Krause, Franz, and Mark Harris. 2021 *Delta Life: Exploring Dynamic Environments Where Rivers Meet the Sea.* New York: Berghahn Books.

Krause, Franz, and Veronica Strang. 2016 "Thinking Relationships Through Water." *Society and Natural Resources*, 29(6): 633–638.

Lahiri-Dutt, Kuntala, and Gopa Samanta. 2013 *Dancing with the River: People and Life on the Chars of South Asia.* New Haven: Yale University Press.

Langston, Nancy. 2003 *Where Land and Water Meet: A Western Landscape Transformed.* Seattle: University of Washington Press.

Leopold, Aldo. 1968 *Sand Country Almanac.* New York: Oxford University Press.

Lévis-Strauss, Claude. 2021 *Wild Thought.* Chicago: The University of Chicago Press.

Lewis, Martha Wells, and Kären Wigen 1997 *The Myth of Continents: A Critique of Metageography.* Berkeley: University of California Press.

Linton, James, and Jessica Budds. 2014 "The Hydrosocial Cycle: Defining and Mobilizing a Relational-Dialectical Approach to Water." *Geoforum*, 57 (November): 170–180.

Mawani, Renisa. 2018 *Oceans of Law: The Komagata Maru and Jurisdiction in the Time of Empire.* Durham: Duke University Press.

Mbembe, Achille. 2021 *Out of the Dark Night: Essays on Decolonization.* New York: Columbia University Press.

Nadarajah, Yaso, and Adam Grydehøj. 2016 "Island Decolonization: Island Studies as a Decolonial Project." *Island Studies Journal* 11(2): 437–446.

Neal, William J., Orrin H. Pilkey, and Joseph T. Kelley. 2007 *Atlantic Coast Beaches: A Guide to Ripples, Dunes, and Other Natural Features of the Seashore.* Missoula: Mountain Press Publishing Company.

Pastore, Christopher L. 2014 *Between Land and Sea: The Atlantic Coast and the Transformation of New England.* Cambridge: Harvard University Press.

Pearson, Michael. 2003 *The Indian Ocean.* New York: Routledge.

Pearson, Michael. 2006 "Littoral Society: The Concept and the Problems." *Journal of World History* 17(4): 353–373.

Pugh, Jonathan. 2018 "Relationality and Island Studies in the Anthropocene." *Island Studies Journal* 13(2): 93–110.

Raadik Cottrell, J., and Stuart P. Cottrell. 2020 "In Spaces in Between—From Recollections to Nostalgia: Discourses of Bridge and Island Place." *Island Studies Journal* 15(2): 273–290.

Roberts, Brian Russell. 2021 *Borderwaters: Amid the Archipelagic States of America.* Durham: Duke University Press.

Roberts, Brian Russell, and Michelle Stephens. Eds. 2017 *Archipelagic American Studies.* Durham: Duke University Press.

Rozwasowksi, Helen M. 2018 *Vast Expanses: A History of the Oceans.* London: Reaktion Books.

Samuelson, Meg. 2017 "Coastal Form: Amphibian Positions, Wider Worlds, and Planetary Horizons on the African Indian Ocean Littoral." *Comparative Literature* 69 (1):16–24.

Schama, Simon. 1995 *Landscape and Memory*. New York: Vintage.

Schneider, Rebecca. 2020 "This Shoal Which Is Not One: Island Studies, Performance Studies, and Africans Who Fly." *Island Studies Journal* 15(2): 201–218. https://doi.org/10.24043/isj.135

Sen, Sudipta. 2019 *Ganges: The Many Pasts of an Indian River*. New Haven: Yale University Press.

Shiva, Vandana. 2002 *Water Wars: Privatization, Pollution, and Profit*. Cambridge: South End Press.

Shiva, Vandana. 2015 *Soil Not Oil: Environmental Justice in an Age of Climate Crisis*. Berkeley: North Atlantic Books.

Starolsielski, Nicole. 2015 *The Undersea Network*. Durham: Duke University Press.

Steingberg, Ted. 2014 *Gotham Unbound: The Ecological History of Greater New York*. New York: Simon and Schuster.

Subramaniam, Ajantha. 2009 *Shorelines: Space and Rights in South India*. Stanford: Stanford University Press.

Uddin, Nasir. 2020 *The Rohingya: An Ethnography of 'Subhuman' Life*. New Delhi: Oxford University Press.

Urbina, Ian. 2019 *The Outlaw Ocean: Journeys Across the Last Untamed Frontier*. New York: Knopf.

1 Kerala Coast and the Environmental Ethics of Precarity

May Joseph

The Climate Is Personal

"Who has known the Ocean?" asked Rachel Carson in 1937 a time when India was still colonized by the British, and Indians had no rights over their coastline or their oceans (Carson 1937, 55; Sen 2002). Storm surge and precipitation have altered what we know about our oceanic ecologies today.[1] There is an urgency to write about environments one has experienced through one's lifetime, to chronicle what Sonja Boon calls "what the oceans remember" (Boon 2019). Growing up along the Kerala coastline, Carson's question resonated with me. Now, as the tidelines of the Kerala shore extend into the hardscape of cities, impacting human habitats alongside the nonhuman and the more-than-human, I find myself asking Carson's question a different way. How is climate change impacting the Kerala coast?[2]

The ocean is a dominant but little-understood site in South Asian maritime history. Commercial and trading accounts of the Indian Ocean and the Bay of Bengal have bookended coastal life in India from a historical perspective (Pearson 2003).[3] Our anthropocenic era of climate engagement necessitates a phenomenological inquiry into coastal environments (Srinivas et al. 2020).[4] Drawing on ocean historian Jennifer Telesca's adage that the best climate writing emerges from writers who have personal stakes in their material, I wade into the *Terra Aqua* from Kochi to Alleppey to Kollam, Varkala, and Vizhinjam along the Malabar coast. Palm-fringed: it is a string of barrier islands forming a serrated nearshore archipelagic intercoastal system of waterways spanning three lakes, Vembanad, Kayamkulam, and Ashtamudi Kayal. This is the largest intercoastal waterbody in India.

For all the spectacular beauty India's extensive coastline boasts, it occupies a pariah place in the ontology of dominant Indian place making. There is an elaborate environmental vernacular of water

DOI: 10.4324/9781003282471-2

in Hindu philosophy, but the coast in contrast has historically been delineated as "barbarian" in the Vedic texts—for being "anupa" or marshy (Zimmermann 1982, 7, 25). It is a place of inclement sensations that is also a place of non-Brahmins and strangers. This cultural hier-archization of India's pelagic habitats as peripheral geographies is deeply entrenched in the national psyche (Subramaniam 2009). The coast is considered barbarian in the Laws of Manu. It is a place where those who are non-Brahmin are to be relegated (Zimmermann 1982). In tension with this dominant strain in Indian environmental philosophy, the South Indian 1500-year-old classical text the *Thirukkural* by the Tamil poet Thiruvalluvar emphasizes water habitats, "Twofold waters, fertile hills with rivers/And forts are a country's limbs" (Thiruvalluvar 2009, 151). The coast in the *Thirukkural* is essential to a successful society "Clear water, open land, mounts and forests/With cool shade form a good fortress" (152). The *Thirukkural* is the earliest precolonial text on environmental precarity emerging from India's southern coast. Through sections such as the "Excellence of Rain" and "Assessing the Place" the *Kural* as it is referred to is a compendium of environmental ethics with attention to climate, rain, oceans, and the nonhuman. "Even the boundless sea will shrink in nature/If the rainy clouds fail to shower" writes Thiruvalluvar presciently (5). "The vast ocean-bound earth suffers/From famine if the sky falls" (4). He notes, "It is rain that ruins and again it is rain/That lifts the ruined to gain." His words presciently anticipate our current climate crisis. Again, Thiruvalluvar writes "Without water life cannot sustain/Nor can virtue without rain". Finally, he observes "As the unfailing rain sustains the world/It is deemed a divine food" (4).

Thiruvalluvar's rich repository of climate wisdom sheds light on South Indian environmental ethics. The *Kural* addresses India's coastal ecol-ogies with great precision in the chapter "Fort". "Clear water, open land, mounts and forests/With cool shade form a good fortress" (Thiruvalluvar 2009, 152). Despite this wide-angled environmental approach to addressing the impact of water on India's coasts, it is the caste-bound prohibitions embedded in the *Laws of Manu* that have shaped Indian environmental thought prior to and during European colonialism to the present. The Portuguese built churches along the shoreline of the Kerala coast solidifying caste prejudice against the coasts as a place of strangers, while the Dutch further endorsed the anti-coast bias as they established their presence along the intercoastal waterway in the Malabar during the 17th century (Gupta 2018; Heniger 1986).

Living in my mother's ancestral home by the sea along this old Portuguese and Dutch maritime route in Kollam, I am immersed in the murky, muddy, and harsh saltwater terrains that are sustained by brackish marine ecologies (Koshy 1989). It is a region of coastal livelihoods in upheaval. The eroding shoreline presents the unfair burdens that regions of the global South have come to bear as a result of the abnegation of mitigation responsibility of the global North. Watching the Malabar wash away is an experience that can best be described as a condition of submergence amid land and sea. It is a wrenching symptom of what lies ahead for low-lying regions around the world.[5]

Toward a Blue and Brown History

The ancient *Thirukkural* is an enlightening manual on regional climate thought, but it is in Carson's work I find the most resonance in relating to the Kerala coast in all its damp, moldy hydro-ecologies of perennial precipitation and flooding. Carson demands attention to both the edges of the sea as well as what lies under the sea (Carson 1998). Of the *Terra Aqua* Carson writes "The edge of the sea is a strange and beautiful place" (Carson 1998, 1). If you live along the Malabar coast, it is indeed a terrible beauty, at once majestic and catastrophic. Carson's invitation to dive into the undercurrents of the *Terra Aqua* is a methodological opening to intersectional and critical thinking linking the personal to the science of climate change in understanding the history of our changing climacteric realities. "The shore has a dual nature, changing with the swing of the tides, belonging now to the land, now to the sea" writes Carson

> Only the most hardy and adaptable can survive in a region so mutable, yet the area between the tide lines is crowded with plants and animals. In this difficult world of the shore, life displays its enormous toughness and vitality by occupying almost every conceivable niche
>
> (Carson 1998, 1)

This amorphous in-between topography that constitutes much of Kerala's wet landscapes has always been terraqueous, Campling and Colàs (2021) would argue. It is a topography where people have "walked on water" as Dilip Menon puts it (Menon 2020). But claiming the shore's blurry coastline is an historical process (Iyer 2007). It takes the historical conditions of sovereignty and citizenship to occupy Carson's "sunless sea" and explore her "encircling sea" (Carson 1989).

Carson's question as to who has known the ocean raises uneasy, differing histories of knowing that the seas themselves raise in the era of decolonization. The free sea of Hugo Grotius was only free for colonizing maritime powers, of which Portugal's "Mar Portugues" or the Portuguese Sea was a metaphor (Grotius 2004; Pessoa 1998; Subramanyam 1998).[6] The space between land and sea was always a violent and brutal space of world dominion (Schmitt 2015). In the case of South Asia, the coast was a site of prohibition, deprivation, and militarization during the era of colonial rule (Amrith 2018; Bishara 2017). The British forbade Indians from accessing the coast to protect their extractive interests in salt exports (Sen 2018).[7] The subalterns who knew "how to read water" worked and lived along the coast of colonial India under oppressive conditions of slavery and indentured labor (Gooley 2016; Shankar 2022a). Hence, the ecological history of the South Asian coast as a place of a coming community is only a recent post-independence process that is being charted out in the throes of the climate crisis (Agamben 1993; Arunachalam 2004). This is a scaffolded history of coastal knowledge lost and washed ashore alongside the muddy traces of the *Terra Aqua* (Shankar 2022b).

There are many points from which one can begin to read India's cultural reclamation of its shores from the scourge of Portuguese, Dutch, and British colonialism. The most evocative historic moment is Mahatma Gandhi's defining Dandi march or Salt Satyagraha—the long walk from his ashram in Sabarmati to the sea-town of Dandi on the western Indian shoreline in Surat. The 240-mile long walk from inland to coast in 1930 marked the beginning of a new historical discourse of decolonizing the marshy peripheries of the colonial state. In protest against the British salt taxes and laws prohibiting Indians from collecting and producing salt from their own shores, Gandhi walked to the coast, immersed himself in the sea and picked a muddy handful of salt, thus breaking British colonial law (Kumar 2019, 99). Gandhi's historic actions were followed by similar acts of civil disobedience against the salt laws along the coasts of India.[8] In Kerala, K. Kelappan led a group of 33 satyagrahis from Calicut to Payyannur in North Malabar.[9] This radical saltwater action intervention catalyzed a pre-independence rethinking about the importance of coastal communities to the emerging nation's political future. Unfolding as a 24-day walking gesture of civil disobedience or *satyagraha* protesting British occupation, the Dandi march presents a vastly different history of the edge of the sea as a space of ecological and political reclamation to that of the global North at the time. This coastal performance of counter-sovereignty foregrounded the importance of India's *terra aqueous* spaces, where land and water, salt and sweat, human, the less-than-human and non-human ontologies congeal. Along India's western coastline in 1930, the colonized and socially oppressed classes of India's caste system and the materiality of salt and sand merged in a historic interrogation between lands' end, and what was for the colonized subject the unfree sea (Grotius 2004).

Gandhi's historic communion with the sea challenged British hegemony and opened the question as to who owns the shoreline and by extension, the emerging nation-state (Kulke and Rothermund 1986). Gandhi's Salt Satyagraha politicized the marshy ontologies of India's shoreline as an ecological space of anti-colonial materiality and self-realization. The colonial Indian shore with its pillaged salt fields and destroyed coastal ecologies became visible as the place of the new subject of India's anti-colonial history, the coastal subaltern, feet muddy and body submerged across India's vast seashore.[10] The fisherman and salt pan worker, the longshoreman and dispossessed coastal dweller became the new actors of the emerging sovereign nation (Menon 1987).

The historian Romila Thapar notes that the extensive western and eastern coasts of India have not been given their "due recognition, largely because the historical perspective of the sub-continent has been landlocked" (Thapar 2022, 46). Following Independence, the Indian shore as a critical space receded in importance and the heartlands became prioritized with New Delhi as the capital of the new India (Ludden 2002, 75–76). Subsequently, the environmental history of postcolonial India has been a green history of the forest and the sacred grove, while the coasts remained under-theorized (Prasad 2004). This neglect of India's beaches began in the wake of the prohibition of coastal spaces under the British, followed by the militarization of the coast under the Indian Navy in the case of the Kerala coastal waters and the security state. Hence, theorizing Kerala's oceanic and tidal futures has been slow to embrace an environmental blue history, of which Thiruvalluvar's *Thirukkural* is its precursor.

Picking up from Thiruvalluvar's observation that "Tillers will not plough and toil/If the rain is not genial" (Thiruvalluvar 2009, 4), Rohan D'Souza's study of the river ecologies of Eastern India is noteworthy for its critical point that "the phenomenon of colonialism itself has been little explored or explained through its ecological footprint" (D'Souza 2006). His history of India's silt-laden flood waters and the destruction of their ecosystems considers the materialities of embankments, canals, dams, fluvial currents, tidal action, and other flood-vulnerable landscapes that structure the river ecologies impacting coastal India to analyze the extractive economies at work in deltaic hydrology during the colonial era. Flood control is a political project, D'Souza argues (D'Souza 2006, 222). British colonialists distorted traditional inundation patterns for single cash crop cultivation geared toward colonial exports. D'Souza throws into relief the deltaic landscapes of mud, silt, and sludge that characterize much of South Asia's riverine landscapes that feed into the Bay of Bengal to argue that extractive strategies of capitalist colonialism recast deltaic inundations as calamitous events rather than adapting to them as a geomorphological process (D'Souza 2006, 223). These irrigation interventions in turn aggravated hydraulic volatility causing an escalating disequilibrium leading to flood-prevention strategies which have in turn generated the current crisis of water having nowhere to go.

Also attending to the vast silence of India's coasts, Ajantha Subramaniam's study of the Mukkuvar fishers at the tip of India in Kanyakumari, situated between Kerala and Tamil Nadu, presents a volatile and dynamic history of fishing communities who constitute themselves as subjects of rights in relation to the local, the regional, and the state. Subramaniam points out that the people who live and work on the seashores of India are given scant mention in the scholarship of transoceanic trade and disappear as historical subjects of the coast. Disrupting cliches about the "primitive coast" Subramaniam presents fishing communities from the southwestern "fishery coast" as political players shaping the emergent fabric of Indian democracy (Subramaniam 2009, 5). In tension with the agrarian-based ecological thought that propelled post-independence Indian environmentalism (Guha 2000), Subramaniam foregrounds the tenuous spaces of the *terra aqueous* that she documents as predominantly Christian and lower caste in social organization, in ways that have not been previously quantifiable, and as the coasts have swelled and abated in the years following independence. Subramaniam sheds light on the tensions between agrarian inland low-caste groups versus the coastal inheritors of the encounter with the Indian Ocean trade such as the Christians of Kanyakumari's fishing villages. Today, the Indian shore is a highly mediated space where its colonial pasts, post-independence militarized interests, and caste-defined social stratifications have only begun to produce an open engagement with the edges of the sea. This frothy, intertidal coast is the urgent space of a new coastal emergency.

Nervous Archipelago

Life in archipelagic Kochi is increasingly a nervous one today. "Not a blade of grass will be seen/If the sky showers no rain" writes Thiruvalluvar, but in Kochi it is too much of a good thing (Thiruvalluvar 2009, 5). The estuarial ecologies and ocean topography of the city are generating a new archipelagic knowing. It is a shifting environmental understanding nuanced by the intricate web of islands and sandy spits that have held the city together. Wedged between mountains and the sea, Kochi's coast is a misty, swelling, leaking, flooding, sinking and deluging interface of land and water. It is intensely poised between monsoonal excesses of recent years and the attending subsidence of earth and mud from elevated regions toward the coast. Monstrous storms, thundering rains, and cyclonic waves are just a meager roster that living along the Malabar implies. Its porous shorelines are blurring the distinction between ocean and land. Higher up in the rain-drenched hills of the hinterlands, the topography is alarmingly transitioning into new landscapes of ponds, puddles, and muddy swamplands. The last few years have been an onslaught of unprecedented precipitation (Hunt and Menon 2020).[11] Heavy rains for extended periods, and increasingly violent storms that wash away the tenuous coastlines, are forcing a new ethics of the nonhuman. Once again, the words of Thiruvalluvar come to mind: "Rain produces food for all beings in the world/And rain itself serves as food indeed" Thiruvalluvar 2009, 4). While Kerala's atmosphere has always been swampy, muggy, and rainy, the climate is changing faster than the myths can accommodate.

Disappearing Ecologies

As India invests large amounts of military and economic capital into the development of Kochi as one of South Asia's major container port cities, the city's rapid transformation from a sleepy medieval port city into a mega port terminal is tearing apart its once verdant coastal ecologies, which were unwittingly preserved through low impact engagement (Bose and Varughese 2015). The distinctive tropical ecology of coastal wilderness, wetlands, and some of the last biodiversities bearing the flora of colonial incursions, with its exotic blend of Dutch, Portuguese, Brazilian, African, and East Asian plant species, have been entirely decimated on the islands of Bolghatty, Vallarpadam, Willingdon, and Vipin (Bernard 1995). Some of the last thriving mangroves of the Malabar region have been filled in without environmental oversight, damning the flow of the Periyar river and its tributaries along the undulating coast (Shanij et al. 2015). An added climate impact factor is the invisible but looming threat that the intricate system of over 80 dams networked across Kerala's 44 river systems will reach unsustainable levels leading to a calamity of unimaginable proportions along the pathway of the many dams.[12] Unregulated development, land reclamation, unsanctioned sand mining, and the aggressive erosion of coastal wetlands have depleted the Kerala coast of its protective habitat.[13] The shoreline is open to the tumultuous sea, bereft of its multilayered topography of foliage, coastal gradations, and mud ecologies. The port with its mega development projects is situated under the shadow of the oldest and largest dam in Kerala, the Mullaperiyar dam in the Idukki district.

Archipelagic Awakening

The warming sea is forcing the city of Kochi into an awakening that it is an archipelago. Traditional understandings of its environmental geographies treat the city as a series of independent islands (Chandler and Pugh 2021; Roberts 2021). With the new bridge and boat infrastructure, a contemporary archipelagic consciousness has emerged in Kochi. Floods, inundation, and storm surges have activated a floating ecology of uncertainty along Kerala's waterlogged terrains—both along the mountains and its coasts.[14] Suddenly, the island structure of the somnolescent barrier islands with their shifting mud banks and sand strips has acquired a new vulnerability. What was once an informal knowledge of social life along the barrier islands of the Malabar has now become burdened by the discourse of adaptation and mitigation. A water-bound life fueled by canoe travel between islands is being drastically reshaped into an intermediary transit hub. As Thiruvalluvar observes in the *Thirukkural*, "Without water, life cannot sustain/Nor can virtue without rain" (Thiruvalluvar 2009, 5). Water shapes life in Kochi, but the torrential rains of Summer 2018 burst open a new watery precarity for the city's future.

The Torrent

August 2018 saw the most violent precipitation over a century in the Kerala region. Within a period of two weeks, the state's dams were flowing at capacity. Inundated without respite, the region's water management technocrats were in a crisis. The question of how much water to release, and how to manage the flow of such large volumes of excess water, failed to register at the national or regional level. Furthermore, the technical quandary that Kerala's largest dam structure—the Mullaperiyar dam complex, is governed by the neighboring state of Tamil Nadu, complicates water management issues. A logistical left over from the colonial British era, Tamil Nadu's decision-making along the Mullaperiyar dam impacts environmental consequences on the Kerala side of the Annamalai region. Consequently, in August of 2018, large swaths of the Kerala population were uninformed of the impending catastrophe of dam overflow which was released by Tamil

Nadu.[15] Like much of India, the informal residential patterns of many hamlets and local enclaves along the path of the dam's surge path meant that breaching the large number of dams could create a major disaster of unimaginable proportions. Imagine the fury and fear of a relentless cataclysm of thundering rains followed by the threat of voluminous water rushing down the mountains from released dams. The irresponsible actions of the Tamil Nadu government's decision to release water from the Mullaperiyar dam without adequate warning and preparation with the Kerala government continue, despite requests from the Kerala government for better communication. Large groups of people on the Kerala side of the dam's pathway live in continuing fear and uncertainty of flooding without warning. How such far-reaching repercussions of impending harm could be handled in a manner from the Tamil Nadu side, where people along the path of the overflow from the Mullaperiyar dam on the Kerala side of the border are just swept away, or entirely displaced—is just one instance of the ethics of shared water governance that is inflected by colonial decision-making within contemporary geographies.

Climate Anomalies

Escalating extreme weather along the Kerala coast over the last two decades has created an accelerated experience of precarity along the Kerala coastline. Beginning with the 2004 tsunami which affected many villages along the coast, an emerging awareness of the ocean's disturbance has percolated into daily lives. Spiritual practices incorporating an environmental consciousness along Kerala's coastal promontories such as the Kreupasanam Marian Shrine in Alleppey suggests an oceanic shift in its visual messaging. A tsunami wave dominates the religious center's central Virgin Mary icon. The Kreupasanam Shrine incorporates narratives of apparitions and miraculous rescues from the catastrophic deluge that occurred as a result of the Sumatra–Andaman earthquake in the Indian Ocean. The Kreupasanam Shrine, which services fishing communities of Alleppey, is just one instance of a new coastal consciousness addressing the local sense of a climate out of balance.

Along the Kochi–Kollam coast, fishermen, laborers, longshoremen, and former seafaring families speak of a difficult ocean whose inconsistencies stem from human hubris. Their accounting of the increasing disruptions to a long history of coastal weather patterns is embedded in a new environmental vulnerability. Living along the shoreline, these communities speak with remarkable lucidity about an increasing ecological forgetting of the sea's delicate balance. They narrate escalating symptoms of human abnegation of environmental understanding rising around them: overfishing, unregulated overbuilding, ocean degradation from garbage dumping, and an increasing toxicity of the waterways upon which they depend for their livelihood.

The painful realities for fishing communities living on the eroding beaches are juxtaposed by the indifference of coastal Keralites who have embraced their topographical wealth without comprehending the significance of its climate vulnerabilities till recently. One young man from Kollam remarks, "For my generation, taking nature seriously is a new concept. We have grown up in a lush landscape without really comprehending the meaning of nature." The young man's observation rings true across the different coastal towns from Kochi to Kollam, whose geographies reflect a class tension between the inland facing middle-class homes and the fishing villages situated along the fragile barrier sand spits that are repeatedly disrupted by escalating floods.

Terror of the Deluge

Wedged between the imposing Annamalai mountains and the Arabian Sea, Kerala's verdant ecology is poised between ocean, wind, mountain, and rain. The coastal settlements comprise a series of barrier islands, settled sand islands, and a vast interconnected series of floating landmasses that have been cultivated over millennia to constitute the archipelagic structure of the Malabar coast. It is an interlaced system of land and sea boundaries, and an intercoastal network of coastal villages, whose exposure to the sea is porous and gaping. The rise of storm surges has led to a vast scenario of climate refugees and an informal culture of disaster relief zones.[16] The long-term viability of many of the existing coastal villages is caught between impending sea levels rising, and the real estate devaluation affecting the fishing communities of the region. The question of resettling coastal peoples away from the coast is an active and urgent one, as impoverished communities in these areas grapple with harsh options.

The barrage of water from the Annamalai mountains in 2018 introduced a new manmade threat from the hills. Sudden, unexpected, and frighteningly furious, the torrent from above onto the Kerala landscape of river towns and villages pushed to the fore the yawning gap between nature and human decision-making. Climate disasters are no longer merely acts of God beyond human control. They are also the terrifying results of infrastructural mismanagement and poor decision-making across the expansive dam technology that has produced modern

Kerala's artificial habitat.[17] Unease about the next gush ghosts the daily lives of people picking up the shreds of their existence along the affected pathways of the 2018 floods from above, caused by human intervention. Trust in science and technology to look out for the public good has been shattered in a disconcerting way. When asked how people affected by the floods are doing, some reply "This is India. Floods come and floods go. Life goes on."

The flip remarks I receive to my searching questions of account-ability, outreach, and salvage, foreground my own anxiety that my mother's home along the shoreline lies in the path of a deluge should an aging dam burst. On one hand, such dark thoughts appear to be irrational and paranoid. "Nothing's going to happen to the dam" is the general approach to living precariously along the coast. On the other hand, as my car weaves its way across destroyed roads, damaged homes, and randomly muddy expanses of earth where a road or house once stood on route from Kochi to Kollam, I realize my worries are perfectly reasonable though unregistered by my relatives, who think of the cas-cade from the hills above as merely floods caused by rains, and therefore, anomalies of climate, rather than the catastrophic decisions of water management. The tenuity of the 80 dams in Kerala with their scenario of increasing rains is undeniable. Political skirmishes between the Tamil Nadu and Kerala governments along with corruption stand between addressing climate disturbances and infrastructural planning along the Malabar Coast. It leaves the region unprepared for what lies ahead. What is to be done? "Pray" my mother suggests, as she doggedly refuses to sell her home and relocate. But I suspect that the gods have left on the first boats. We must deal with the consequences of the Anthropocene with eyes wide open.

Environmental Ethics of Precarity

September 1, 2021, witnessed the most catastrophic flooding of New York City's subways and homes in history. It was an extreme event of flash flood that claimed over 37 lives in New York City. The flash floods on higher ground up in Harlem were as violent and catastrophic as the low-lying basement home tragedies of Queens, New York. Clearly, the calamitous moment of unpreparedness in the face of cli-mate urgency is not just a Global South condition. It foists upon the world an environmental ethics of the dammed (D'Souza 2006). What can we do in the face of imminent extreme climate disturbances of unexpected severity? The tragedy of New York was the fact that there was nowhere for the water to go. The vulnerability of Kerala with its reclaimed land and buried water sources covered in concrete bears the same ecological disruptions that there is nowhere for the water to go.

In the face of escalating climate repercussions, Kerala, like New York, must contend with rapacious overdevelopment without attention to low-lying areas and the aggressive destruction of Kerala's coastal ecologies such as swamps mangroves and island biodiversity. This has left the coast open to the elements. What is alluring and beautiful about Kerala is also what makes it fragile; its delicate ecosystem of mud, rock, stone, and brick that remains open to the sea.[18] In the absence of major infrastructural investment bolstering Kerala's interwoven coastal and river systems, the future of a muddier, waterlogged, and sinking coast is unfolding already. What makes this scenario deeply troubling is the reality that some of the most vulnerable populations of the region live along the coast. They own their homes at the interface of land and sea where the oceans have already encroached their meager properties (Subramaniam 2009). But moving is not an option for many of these simple fishing communities who have eked a fragile existence on the margins of *Terra Aqua*.

Bodhisattvas by the Sea

As I finish this essay on a frosty January morning in New York, The National Public Radio announces the grim news that 2021 was the warmest year on record in the history of the planet. The news that President Biden's climate policies will not be able to pass Congress's anti-climate contingent, bears heavily on the ethical imperative behind thinking and writing about planetary futures from the space of the United States, responsible for a quarter of the world's carbon emissions while being less than 5% of the world's population.[19] I close this essay by citing what Lindsay Bremner calls "muddy logics"[20] through the eyes of one old lady who lives alone by the sea in Kollam, my nearly 90-year-old mother. I have come to understand my mother and my extended family living along the shoreline of the Malabar as stewards of the coast. Their historical memory of what the coast used to be, how it has morphed, and the terrifying reality of their witnessing the transformation of their coast because of climate warming is a partial answer to Rachel Carson's imperative to know the ocean. "I remember we had to travel the whole day and night by *vallam* (wooden canoe) from Kollam to Ernakulam for my sister's wedding in the 1950's" Mother recalls. Precariously situated between the ocean and the eight-fingered Ashtamudi Lake, Mother recounts "We used to sleep in the *vallam* with sheets at night. Now, people fear *vallams* because of the climate." she says. "So many accidents due to the weather." Mother says. "The rains nowadays are year round, not like before, when there was a clear monsoon season". Mother describes the recent deluge of violent rain during Fall 2021 as frightening. In all her life along the Kerala coast, she says she has not witnessed such a torrent of water from the sky, ocean, and hills converge. "The weather is out of balance." Mother observes. "We have exploited mother nature. The world is being punished for its greed".

Speaking to Mother in the frightening intensity of Hurricane Ida in New York City in 2021 as it ravages basements, neighborhoods, subways, and infrastructure is distressing. As in India, it is the economically marginal communities of New York who are worst affected by a city planning system still unprepared for extreme weather events such as a 5-inch precipitation within an hour.[21] In Kerala, inundation, flooding, mudslides, sinkholes, catastrophic washing away of infrastructure, all present a cumulative scenario of perpetual climate anxiety. One is constantly in an onslaught of weather—heat , rain, loss of biodiversity, storms. This intensity manifests as gaps in stone embankments washed away like a savaged gash. Large boulders had disappeared along the shore from the last great storm of 2018 at Ayiramthengu and Chavara when I visited the storm-ravaged sites. The sea lazily swept into dwelling areas of the tiny settlements while people still awaited state assistance to restore the damaged coastal buffering infrastructure.

Seasoned coastal communities along the Kochi–Alleppey–Kollam coastal road stoically weather out the calamities to their families by burying their dead and living with extended relatives till it becomes possible to return to their seawater deluged homes. These coastal communities resolutely pick up their lives yet again, once the grief, the loss, and the anger ebb enough to allow them the energy to rebuild their destroyed sandbar homes, juice kiosks, and fishing businesses. They coexist in a relationality with mold, the damp, and the threat of subnature (Gissen 2007).

Terra Aqua is perhaps the most widespread living ecology linking coastal regions around the world impacted by storm surges (Goodell 2018). COP 26 has demonstrated the hypocrisies at play in the deadly game of expendable geographies where entire regions will be unable to adapt to the impending effects of climate change (Shiva 2015). As the Intergovernmental Panel on Climate Change (IPCC) has laid out in 2022, the assessment of the impact of climate in South Asia is dire.[22] From ocean acidification and the acceleration of coastal storms to water stress and irreversible loss to biodiversity, South Asia is precariously poised to encounter climate impact. The IPCC report underscores the point that based on what the people from the global South know about the ocean. The world (not just India and China) needs to implement aggressive mitigation, alongside adaptation, to stem the impending global scenario of a failed ethics of planetarity.

Notes

1 I want to thank Sudipta Sen for opening up the third space of the *Terra Aqua*. This essay emerged out of conversations with Sudipta Sen, Pamila Gupta, Jennifer Telesca, Carl Zimring, Neelima Jeychandran, Viju James, Smriti Srinivas, Arup Chatterjee, Pedro Manuel Pombo, David Ludden, Pius Malekandathil, Brian Russell Roberts, and Nitya Jacob. Rohan D'Souza, Dilip Menon, Devika Shankhar, Charne Lavary, and Lindsay Bremner further helped me understand the swamps of the Malabar. Many thanks to Sudipta Sen, El Glasberg, and Devika Shankar for editorial suggestions. I dedicate this essay to my mother who lives on Ashtamudi Lake.
2 Thank you to Carl Zimring and Jennifer Telesca for sharing research on oceans (Zimring and Corey 2021; Telesca 2020).
3 See also Bose, *A Hundred Horizons*; Desai, *Commerce with the Universe*; Gupta et al. eds. *Eyes across the Water*; Prange, *Monsoon Islam*; Lavary, *Writing Ocean Worlds*; DeLoughrey, *Routes and Roots*.
4 See also Mewani, *Across Oceans of Law*; Joseph, *Sea Log*; Bindra ed. *Voices in the Wilderness*; Rahul Mehrotra and Felipe Vera, *Kumbh Mela*; Samuelson and Lavery, "The Oceanic South".
5 See https://report.ipcc.ch/ar6wg2/pdf/IPCC_AR6_WG11_FinalDraft_Ch apter10.pdf
6 See also Mathews et al. *The* Portuguese; *Journal of The Vasco Da Gama Research Institute of* Cochin; Pagden, *Lords of All the World*.
7 Sudipta Sen. *Ganges: The Many Pasts of an Indian River* (Yale University Press, 2018).
8 A. Sreedhara Menon, *Political History of Modern Kerala* (Kottayam, Kerala: DC Press, 1987), 73.
9 See Rajmohan Unnithan, "Payyanur's Gandhi connection" *The Hindu*, February 24, 2020.
10 See Pulapre Balakrishnan, "They Lived to tell the tale: Revisiting the Moplah Rebellion of 1921" *The Hindu*, August 27, 2021.
11 See also V. Mishra, S. Aaadhar, H. Shah, R. Kumar, D. R. Pattanaik, and A. D. Tiwari, "The Kerala Flood of 2018: Combined Impact of Extreme Rainfall and Reservoir Storage," *Hydrol. Earth Syst. Sci.* Discuss. [preprint], https://doi.org/10.5194/hess-2018-480
12 See A. Latha and Manju Vasudevan, "State of India's River-Kerala", *India River Week*, Kerala, 2016.
13 See Justin Mathew, "Port Building and Urban Modernity: Cochin, 1920-45" (Bose and Varughese 2015).
14 See K. A. Shahji, "Kerala Floods Show State Must Reckon with Climate Change Urgently" *The Huffington Post,* Aug. 13, 2019.

15 See "Don't Open Mullaperiyar Dam at Night: Kerala to Tamil Nadu", *All India Press Trust of India,* December 2, 2021.

16 See Haritha John, "Kerala's climate refugees increase as sea eats into coast", *The Week Magazine,* June 26, 2018.

17 See Dhinesh Kallungal, "Kerala receives record rainfall in century" *The New Indian Express,* November 16, 2021.

18 See the report by the Hindustan Times Correspondent, "Kerala surpasses 2018 tally, records heaviest rainfall in 6 decades", *Hindustan Times,* Nov. 26, 2021.

19 *Union of Concerned Scientists,* "Each Country's Share of CO2 Emissions" July 16, 2008. Updates Jan 14, 2022. ucsusa.org

20 Lindsay Bremner, "Muddy Logics". In: Sheppard, L. and Przybylski, M. (eds.) *Bracket 3: At Extremes.* (Barcelona Actar, 199–206).

21 Signe Nielsen Talk, Pratt Institute, Fall 2021.

22 https://report.ipcc.ch/ar6wg2/pdf/IPCC_AR6_WG11_FinalDraft_Chapter10.pdf

References

Agamben, Giorgio. 1993 *The Coming Community.* Minneapolis: University of Minnesota Press.

Amrith, Sunil. 2018 *Unruly Waters: How Rains, Rivers, Coasts, and Seas Have Shaped Asia's History.* New York: Basic Books.

Arunachalam, B. 2004 *Mumbai by the Sea.* Mumbai: Maritime History Society.

Bernard, K. L. 1995 *History of Fort Cochin.* Cochin: Holy Family Off-Set.

Bindra, Prerna Singh. 2010 *Voices in the Wilderness: Contemporary Wildlife Writings.* Delhi: Rupa.

Bishara, Fahad Ahmad. 2017 *A Sea of Debt: Law and Economic Life in the Western Indian Ocean, 1780-1950.* London: Cambridge University Press.

Boon, Sonja. 2019 *What the Oceans Remember.* Waterloo: Wilfred Laurier Press.Bose, Satheese Chandra and Shiju Sam Varughese eds. 2015 *Kerala Modernity: Ideas, Spaces and Practices in Transition.* New Delhi: Orient Black Swan.

Bose, Sugata. 2006 *A Hundred Horizons: The Indian Ocean in the Age of Global Empire.* Cambridge: Harvard University Press.

Bremner, Lindsay. 2015 Muddy Logics. In *Bracket 3: At Extremes*, edited by L. Sheppard and M. Przybylski. Barcelona: Actar, 199–206.

Campling, Liam and Alejandro Colàs. 2021 *Capitalism and the Sea: The Maritime Factor in the Making of the Modern World.* New York: Verso.

Carson, Rachel. 1937 Undersea. *Atlantic Monthly* 160(3): 55–67.

Carson, Rachel. 1941 *Under the Sea-Wind.* London: Penguin.

Carson, Rachel. 1989 *The Sea Around Us.* New York: Oxford University Press.

Carson, Rachel. 1998 *The Edge of the Sea.* New York: Houghton Mifflin Harcourt.

Chandler, David, and Jonathan Pugh. 2021 *Anthropocene Islands, Entangled Worlds.* London: University of Westminster.DeLoughrey, Elizabeth. 2007

Routes and Roots: Navigating Caribbean and Pacific Island Literatures. Honolulu: University of Hawai'i Press.

Desai, Gaurav. 2013 *Commerce with the Universe: Africa, India, and the Afrasian Imagination.* New York: Columbia University Press.

D'Souza, Rohan. 2006 *Drowned and Dammed: Colonial Capitalism and Flood Control in Eastern India.* New Delhi: Oxford University Press.

Gissen, David. 2007 *Subnature: Architecture's Other Environments.* New York: Princeton Architectural Press.

Goodell, Jeff. 2018 *The Water Will Come: Rising Seas, Sinking Cities, and the Remaking of the Civilized World.* New York: Back Bay Books.

Gooley, Tristan. 2016 *How to Read Water.* London: Hatchette, U.K.

Grotius, Hugo. 2004 *The Free Sea.* Indianapolis: Liberty Fund.

Guha, Ramachandra. 2000 *Environmentalism: A Global History.* New York: Longman.

Guha, Ramachandra. 2014 *Environmentalism: A Global History.* Gurgaon: Penguin.

Gupta, Pamila. 2018 *Portuguese Decolonization in the Indian Ocean World: History and Ethnography.* London: Bloomsbury.

Gupta, Pamila, Isabel Hofmeyr, and Michael Pearson, eds. 2010 *Eyes Across the Water: Navigating the Indian Ocean.* Pretoria: Unisa Press.

Heniger, J. 1986 *Hendrik Adriaan Van Reede Tot Drakenstein (1636-1691) and Hortus Malabaricus: A Contribution to the History of Dutch Colonial Botany.* Boston, Rotterdam: A. A. Balkema.

Hunt, Kieran M. R., and Arathy Menon. 2020 The 2018 Kerala Floods: A Climate Change Perspective. *Climate Dynamics* 54: 2433–2446

Iyer, Ramaswamy R. 2003 *Water: Perspectives, Issues, Concerns.* New Delhi: Sage.

Iyer, Ramaswamy R. 2007 *Towards Water Wisdom: Limits, Justice, Harmony.* London, New Delhi: Sage.

Jamison, Dale, ed. 2001 *A Companion to Environmental Philosophy.* Oxford: Blackwell Publishing.

Joseph, May. 2019 *Sea Log: Indian Ocean to New York.* London, New York: Routledge.

Koshy, M. O. 1989. *The Dutch Power in Kerala (1729-1758).* New Delhi: Mittal Publications.

Kulke, Hermann, and Dietmar Rothermund. 1986. *A History of India.* New York: Dorset Press.

Kumar, Aishwary. 2019 *Radical Equality: Ambedkar, Gandhi, and the Risk of Democracy.* New Delhi: Navayana.

Lavery, Charne. 2021 *Writing Ocean Worlds: Indian Ocean Fiction in English.* Cham: Palgrave.

Ludden, David. 2002 *India and South Asia: A Short History.* Oxford: One World Publications.

Mathew, K. S. 2008 The Portuguese and the Cultural Transformation in Malabar during the Sixteenth and the Seventeenth Centuries. *Journal of the*

Vasco Da Gama Research Institute of Cochin (Bi-annual Research Journal) 1(1): 7–22.

Mathews, K. S., Teotonio R. de Souza, and Pius Malekandathil. 2001 *The Portuguese and the Socio-Cultural Changes in India, 1500-1800*. Pondicherry: MESHAR.

Mawani, Renisa. 2018 *Across Oceans of Law: The Komagata Maru and Jurisdiction in the Time of Empire*. Durham: Duke University Press.

Mehrohtra, Rahul, and Felipe Vera, eds. 2015 *Kumbh Mela: Mapping the Ephemeral Megacity*. Ostfildern: Hatje Cantz.

Menon, Dilip M. 2020 Walking on Water: Globalization and History. *Global Perspectives* 1(1): 1.

Menon, Dilip M., Nishat Zaidi, Simi Malhotra, and Saarah Jappie. 2022 *Ocean as Method: Thinking with the Maritime*. Abingdon, Oxon, New York: Routledge.

Menon, Sreedhara A. 1987 *Political History of Modern Kerala*. Kottayam: DC Books.

Pagden, Anthony. 1995 *Lords of All the World: Ideologies of Empire in Spain, Britain and France c.1500-c.1800*. New Haven: Yale University Press.

Pearson, Michael. 2003 *The Indian Ocean, Seas in History*. London: Routledge.

Pessoa, Fernando. 1998 Portuguese Sea. In *Fernando Pessoa & Co. Selected Poems*, edited and translated from the Portuguese by Richard Zenith. New York: Grove Press.

Prange, Sebastian. 2018 *Monsoon Islam: Trade and Faith on the Medieval Malabar Coast*. London: Cambridge University Press.

Prasad, Archana. 2004 *Environmentalism and the Left: Contemporary Debates and Future Agendas in Tribal Areas*. New Delhi: Leftword Press.

Roberts, Brian Russell. 2021 *Borderwaters: Amid the Archipelagic States of American*. Durham: Duke University Press.

Russell-Wood, A. J .R. 1992 *The Portuguese Empire, 1415-1808: A World on the Move*. Baltimore, London: Johns Hopkins University Press.

Samuelson, Meg, and Charne Lavery. 2019 The Oceanic South. *English Language Notes* 57: 37–50.

Schmitt, Carl. 2015 *Land and Sea*. New York: Telos Press Publishing.

Sen, Sudipta. 2002 *Distant Sovereignty: National Imperialism and the Origins of British India*. London: Routledge.

Sen, Sudipta. 2018 *Ganges: The Many Pasts of an Indian River*. New Haven: Yale University Press.

Shanij, K., B. Peroth, and S. Sivankunju. 2015 Kunhimangalam: The Largest Mangrove in Kerala Needs Immediate Conservation Attention. *Sacon Envis Newsletter-Sarovar Saurabh* 2(2).

Shankar, Devika. 2022a Water, Fish and Property in Colonial India, 1860–1890. *Past and Present* https://doi.org/10.1093/pastj/gtab043

Shankar, Devika. 2022b A Monsoon Miracle: Naming and Knowing the Mudbanks of Kerala. In Sudipta Sen and May Joseph eds. *Terra Aqua: The Amphibious Lifeworlds of Coastal South Asia*. Abingdon, Oxon: Routledge.

Shiva, Vandana. 2015 *Soil Not Oil: Environmental Justice in An Age of Climate Crisis*. Berkeley: North Atlantic Books.

Srinivas, Smriti, Bettina Ng'weno, and Neelima Jeychandran, eds. 2020 *Reimagining Indian Ocean Worlds*. London, New York: Routledge, 2020.

Steinberg, Ted, and Kimberly Peters. 2015 Wet Ontologies, Fluid Spaces: Giving Depth to Volume Through Oceanic Thinking. *Environment and Planning D: Society and Space* 33(2): 247–264 https://doi.org/10.1068/d14148p

Subramaniam, Ajantha. 2009 *Shorelines: Space and Rights in South India*. Stanford: Stanford University Press.

Subramanyam, Sanjay. 1998 *The Career and Legend of Vasco Da Gama*. London: Cambridge University Press.

Telesca, Jennifer. 2020 *Red Gold. The Managed Extinction of the Giant Bluefin Tuna*. Minneapolis: University of Minnesota Press.Thapar, Romila. 2022 *The Penguin History of Early India: From the Origins to AD 1300*. Delhi: Penguin.

Thiruvalluvar. 2009 *Thirukkural*. New Delhi: Rupa.

Zimmermann, Francis. 1982 *The Jungle and the Aroma of Meats: An Ecological Theme in Hindu Medicine*. Delhi: Motilal Banarsidass.

Zimring, Carl, and Steven H. Corey. 2021 *Coastal Metropolis: Environmental Histories of Modern New York City*. Pittsburgh: University of Pittsburgh Press.

2 A Monsoon Miracle

Naming and Knowing the Mudbanks of Malabar[1]

Devika Shankar

I The Malabar Mudbanks

If you pursue the history of ports along Malabar long enough, sooner or later you are bound to encounter references to a peculiar and mysterious phenomenon called the mudbank, a rather mundane sounding term that in the context of India's southwest coast evokes the marvellous. For across parts of the region, the appearance of these mudbanks, or highly sedimented patches of water, is associated with remarkably tranquil seas and the profusion of dead fish during the monsoons, a time when the sea is usually marked by menacingly high waves, making the seasonal occurrence a boon for navigation and fishing alike. While fairly similar mudbanks have been noted in other muddy coasts across the world, especially around the Amazon, what makes their formation in Malabar unusual is their transient and seasonal nature (Mathew 1992). What is more, unlike in other locations, mudbanks in Malabar can also appear in the absence of any apparent river discharge. Along India's southwestern coastline, these distinctive mudbanks have been noted at several locations, but there are four that are especially important because of their proximity to some of the region's most prominent ports. If their proximity to commercial centres has helped draw attention to the mudbanks at these locations, these formations have in turn shaped the development of the ports in their vicinity in fundamental ways. Whether it is in the case of Calicut, Cochin, or Alleppey, therefore, important episodes in the histories of each of these ports have become closely entangled with the appearance and disappearance of these mudbanks.

Take for instance, what is often considered to be the inaugural moment of European colonialism in South Asia—the arrival of Vasco da Gama on the coast of Calicut. In 1498, Da Gama and his crew reached Calicut in May in the midst of menacing squalls, and their ships are said to have stayed in the open sea up until August, a period

DOI: 10.4324/9781003282471-3

that almost perfectly coincides with the monsoon season when the seas around Calicut are extremely choppy.[2] How could the Portuguese have survived in the open sea under such conditions? Several 19th century observers claimed that the only real possibility was that Vasco da Gama had found his way to one of the smaller mudbanks on the coast in Pantalayini Kollam next to Calicut (Logan 1989, 297). Many others however disputed this theory and pointed to the absence of any references to the sea's uncharacteristic tranquillity in contemporary Portuguese accounts. It was in fact only from the 18th century onwards that European travellers would begin to notice and describe the unusually calm waters along parts of the Malabar coast with increasing frequency and wonder.

While it is through the eye that the mudbank's qualities are best perceived, its uniqueness is hard to capture and reproduce visually either through painting or photography. This is largely because its unusual features only become apparent when contrasted with the rest of the coast. What is more, the profusion of mud that gives the phenomenon its name, cannot always be seen, often it can only be felt. So even as it can enthral a person standing on its shores, it can appear quite unremarkable in its visual reproductions. Due to this, once Europeans began to notice the mudbank in the early modern period, it was largely through narrative accounts that they attempted to communicate the uniqueness of the mudbank and its exceptional nature. Through a focus on these narratives, this chapter will highlight the ways in which European and local observers have made sense of the mudbank over the last few centuries. It will simultaneously examine the struggle to describe and represent a phenomenon that straddles and challenges the foundational boundary between land and sea.

II Narratives of Exception

The first-known European description of the mudbank dates to the 18th century. Written by a Scottish merchant and sea captain Alexander Hamilton, this account was subsequently included in a popular compilation of travel accounts called *Pinkertons Collection of Voyages* published a few decades later,[3] It is this account that provides us with one of the most important and influential descriptions of this phenomenon and its properties. Referring to tranquil seas around the port of Alleppey in southern Malabar, Hamilton stated,

> Mudbay is a place that... few can parallel in the world. It lies on the shore of St. Andrea, about half a league out in the sea and is open

to the wide ocean, and has neither island nor bank to break off the force of the billows which come rolling with great violence on all parts of the coast in the south-west monsoon, but on the bank of mud lose themselves in a moment and ships lie on it, as secure as in the best harbor without motion or disturbance.[4]

At a time when European travel accounts sought to represent and comprehend Indian difference through evocations of the marvellous, the exceptionally still waters of Malabar were unsurprisingly represented as a wonder that could not quite be explained in terms of the usual properties of land or water (Nayar 2005). What is more interesting, however, is that in this text the mudbank was identified not as a natural phenomenon but as a place called "Mudbay". One can, however, already see early echoes of the term mudbank which would gain popularity over the following centuries, with Hamilton referring to the "bank of mud" on which ships could lie without being disturbed.

Throughout the 18th century, most English and Dutch accounts would continue to use the terms Mudbay and *Modderbaai* respectively, treating the calm waters as the distinctive feature of a particular place close to Alleppey while simultaneously emphasizing the exceptional nature of the waters in that locality. These accounts would not only reproduce Hamilton's narrative almost verbatim, company officials would also in their correspondence urge their ships to find their way to Mudbay during the monsoons. For instance, in the *Seaman's guide to the navigation of the Indian Ocean and the China Sea*, published in 1867, the author provided directions to Mudbay and emphasized the importance of publicizing its special qualities. Referring to its importance for navigation, he stated,

> I certainly never during the 22 years I have been trading to Bombay had the remotest idea that I could land anywhere on the coast in the bad season, and I believe most of shipmasters are in equal ignorance.[5]

Why did early European observers, who had either encountered the phenomenon first-hand or heard about it from others refer to it as a bay? This was a question raised by several later 19th-century figures with some speculating that the area occupied by the mudbank might have once formed an actual bay with subsequent accretions of land altering the configuration of the coast (Menon 1982, 26–27). But a closer look at the name and the context of its appearance also suggests other possibilities.

The practice of naming, as several scholars have argued, can be an important object of study not only because naming often constitutes the object that it purports to represent but also because a close examination of the choice of names can be revelatory in its own right. In the *Road to Botany Bay* Paul Carter, for instance, closely tracked the names that Captain Cook and other early explorers applied to various places in Australia to argue that these names were not accidental, but were in fact reflective of the journeys through which these places were "found". According to him, therefore, place names need to be situated and understood in the context of the texts within which they first appeared. In this respect, it is not just proper names that needed to be critically investigated but also class names like "hills" and "rivers". "Geographical class names", as Carter evocatively states, "created a difference that made a difference. They rendered the world visible, bringing it within the horizon of discourse" (Carter 1987, 51). Others have also pointed out that the development of natural history in the early modern period had contributed to an interest in classification, with novel natural formations and forces being absorbed into familiar categories. This impulse could, as Carter shows, lead to the constant search for analogies with the strange or unique being sought to be rendered in familiar terms. In the process, terms like bays and harbour would be applied to places because these provided the closest analogues for the novelties encountered.

Seen in this light, we need not assume that the use of the term Mudbay indicated the actual existence of a bay at some historical juncture. Instead, the term could equally reflect an uneasy attempt to mark out the specificity of a place that could not be fully enfolded in any existing categories. Indeed, it has been argued that this search of analogies was one of the characteristic and distinctive features of Hamilton's travel account in which the term Mudbay first appears (Markley 2007). A close examination of this travel account has in fact prompted one scholar to remark that Hamilton was animated by a desire to highlight the geographical specificities of the regions he visited while also making them more intelligible to an European audience through the use of analogy. It is probable, then, that Hamilton might have used the term Mudbay because the term highlighted two of its most salient features. One was its association with mud and the other was its similarity to a bay not necessarily in form, but in function. Bays are after all known for providing shelter from winds and waves for fishing and navigation which is exactly what Malabar's mud formations did. Much like a bay, then, the muddy waters next to Alleppey were a boon for fishing communities and sailors, perhaps explaining why early European travellers including Hamilton referred to this spot as a bay. But as observers began to notice similar areas in other parts of the coast, it became clear that the object of discussion wasn't just a place but a transient and mobile natural phenomenon. In such a context the inadequacy and inappropriateness of the term "Mudbay" became all too apparent, with the term slowly being replaced by others indicating a natural phenomenon rather than a location.

While throughout the 19th century observers would use multiple names to refer to this phenomenon, including mudflats, by the final decades of the century the term mudbank would displace all others to emerge as *the* term for the seasonal appearance of tranquil waters around parts of the Malabar coast. The stabilization of the category would of course accompany and reinforce its emergence as a scientific object, and the late 19th century would witness a number of important investigations into the nature of these mysterious formations. While later observers would express surprise at the fact that the phenomenon was known through the geographical designation "Mudbay" for more than a century, the term mudbank was also far from perfect. This was not only because the term has been used for a wide variety of mud formations across the world but also because the use of the term "bank" seems to suggest the existence of a solid mound of mud. But around Malabar that is often not the case. The mud remains suspended in the water dampening its waves, often without forming a solid mass around it. So while in other parts of the world "mudbanks" were often navigational hazards that could ground ships, in Malabar they created an almost miraculous anchorage. In fact, in Alexander Hamilton's original account "the bank of mud" on which ships could seek shelter was explicitly contrasted with these more well-known solid mudbanks, with the author stating that what was remarkable about Mudbay was that its waters were calm despite the lack of an island or bank to break the waves. The mudbanks of Malabar also differed in crucial aspects from wave-dampening mudbanks in other parts of the world, including the Amazon and some parts of China, where they occurred very close to river mouths, rendering the waters so muddy that they posed a danger to navigation. Christopher Columbus for instance is said to have noted the difficulties posed by such mudbanks in his journal on his voyage to the New World (Davies 1953; Parizanganeh 2008). But in some parts of Malabar, mudbanks appeared far from river mouths making the source of mud a mystery. What is more, this mud had the peculiar ability to dampen waves without simultaneously disabling ships. Those trying to describe the phenomenon in Malabar were therefore continually hard pressed to find words to represent and describe a phenomenon that blurred the ontological separation between land and water. This struggle illustrates how riparian areas and formations often fall outside existing scientific and linguistic categories.[6] The foundational land–water binary is so deeply embedded in language that we have to strain our vocabulary to find words to describe places and things that sit uneasily somewhere between the two. While so much about the mudbanks of Malabar still remains a mystery, what is clear is that it is generated and shaped by a wide range of terrestrial, fluvial, and marine processes. Rains, cyclones, earthquakes, sedimentation, currents all seem to play a role in both its formation and its movement.[7]

III Describing Difference

If the struggle to find an appropriate term for this occurrence in English highlights some of the challenges posed by the mudbank's hybrid ontology, then its treatment among local communities in Malayalam is also illuminating in its own way. Like the word mudbank, the local term *chaakara* seems to have emerged out of the combination of two existing words, *chaava* or *chatha* meaning dead and *kara* meaning shore. The term, therefore, according to several scholars, literally just means "dead shore" (Mathew and Gopinathan 2000). Since for local communities the arrival of the *chaakara* is most closely associated with a bounty of fish, some have also pointed out that the term has now come to denote any large haul of fish.[8] If the European terms attempted to tackle the phenomenon's ontology, the local term seems to be more concerned with its phenomenological effects. What it *appears as* rather than simply what it *is*. At the same time, ethnographic work has highlighted the ways in which the *chaakara* has been personified in local discourse with the seasonal phenomenon being understood as a divine visitation, one that can be affected by a variety of natural forces and human interventions.[9] Such personification itself, as Gotz Hoppe points out in his work on the Kerala coast, is a "form of ontological metaphor which may help in organizing knowledge about an otherwise poorly understood environment" (Hoeppe 2004, 247).

It is hard to say just when the term *chaakara* came to be used to designate tranquil waters off the Malabar coast. In fact it doesn't appear in the important Malayalam English dictionaries of the 19th century.[10] In colonial sources too, while the local community's familiarity with the phenomenon is acknowledged and discussed, its local name is not mentioned.[11] The term's broader usage beyond fishing communities seems to have in fact only occurred in the 20th century following the publication of Thakazhi Sivasankara Pillai's Malayalam novel *Chemmeen* and the subsequent release of its cinematic adaptation, centred on the lives of fishing communities and their relationship with the volatile sea. Both the novel and the film became instant classics, with Thakazhi's novel eventually being translated into more than 50 languages and becoming the first Malayalam novel to win the Sahitya Akademi Award in 1957. The film too proved to be immensely popular, winning a national award and being screened at various important international film festivals. The appearance of the mudbank occupies an important position in the novel and the film, an importance reflected by the fact that the film even features a song centred on the phenomenon. It is these literary and cinematic renderings that appear to have popularized the term in Malayalam in the last few decades, helping broaden its appeal and significance outside its immediate coastal context.

The slippery nature of the mudbank is reflected not only in the struggle to name the phenomenon but also in the enduring inability to explain it. Since it was the mudbank's ability to facilitate navigation by tranquillizing monsoonal waves that was most critical as far as the British were concerned, it was this aspect of the phenomenon that drew the most sustained attention during the period of colonial rule. Some attributed this characteristic feature to the action of soft mud at the bottom of the sea, which when "stirred up" by a heavy ocean swell had a calming effect on the waves above, others claimed that the soft mud was brought to the ocean by a subterranean stream or a succession of such streams, which during the monsoons, pushed mud out into the ocean through the backwaters.[12] These were just two of the many theories that were put forward to try and explain mudbanks at a time when the administration was only beginning to recognize and appreciate their tremendous significance. In the absence of systematic scientific enquiries to explain the unique action of mud around the coast, however, much of the speculation surrounding the phenomenon emerged out of piecemeal investigations carried out by various civil and military functionaries in the region.

In 1870, for instance, the commercial agent of Travancore noted that when pipes were being sunk into the ground close to the beach at Alleppey, a large concentration of mud was encountered at a depth of around 80 feet (Robertson 1875; Silas 1984). This mud, he observed, was of the same texture and consistency as that in the mudbanks, causing him to conclude that it was this very subterranean mud that was pushed out into the sea during the monsoons when the rivers and backwaters of the region begin to swell. The unique nature of the mud, especially its oily consistency meanwhile, led others to speculate about the presence of oil in the region. Still others claimed that the mudbanks had something to do with one of the coast's other unique features, its laterite soil (Dinesh Kumar 2016). Over the course of the 20th century, as important infrastructure development projects were attempted and executed along the coastline, the effects of such interventions on the dynamism of the mudbanks also began to be studied closely. Despite these considerable efforts however, such investigations proved to be inconclusive. The mobile mudbanks that had for long been considered as assets for navigation were now increasingly seen as threats to infrastructure proliferating along the coast.

While colonial scientists were largely interested in analysing and explaining their tranquillizing effects, the most striking and useful feature of mudbanks for local communities was their close association with fishing. Though this attribute of the mudbank was neglected in colonial studies, in the post-independence period a number of scientists began to turn their attention to the mudbank's remarkable ability to attract large shoals of fish. Here too, a number of theories have been put forward on the basis of close studies of the composition of the mudbanks. Some have concluded that the chemical composition of mudbanks serves to attract fish while others have attributed the phenomenon's association with fish simply to the fact that during the monsoons, fishing is confined to those parts of the sea tranquilized by mudbanks (Dinesh Kumar 2016, 38–41). Today, these seasonal formations have not only become more erratic, but their characteristic features have also become less striking. The waters are less calm, while more importantly, for local communities that depend on these formations for their livelihoods, they are now associated with less fish. Infrastructural development in these areas and climate change have of course been cited as major reasons for these changes even though scientists are still not entirely sure how these mudbanks form and why.

IV Conclusion

"Mud" is often defined as earth that has been mixed with water. Along coastlines, therefore, mud that usually refers to silt and clay can be regarded as a classic *terraqueous* substance: one that through its very hybrid nature illustrates the impossibility of separating land and water. Malabar's famous mudbanks provide an even more striking example of the inadequacy of rigid classifications along the coast. As a seasonal, transient, and mobile phenomenon, the mudbank has consistently confounded all attempts to stabilize it; not just physically, but also in language. The struggle to name and comprehend the mudbank is, I have argued, a good reflection of the representational challenges involving an object that is hard to define and illustrate using conventional categories.

If the term mudbank expresses the difficulty, or rather the impossibility, of translating the phenomenon's ambiguous and amphibious materiality into words, the struggle to explain it reveals the extent to which the land–water boundary is so fundamental to the way we usually understand the world. In such a context, a phenomenon that can't be reduced to either can only really be understood as a marvel or a miracle.

Notes

1 I would like to thank Noujas V for helping me access and understand scientific literature relating to the mudbank as well as May Joseph and Sudipta Sen for all their help in completing this chapter.
2 For an account of this moment of arrival, see Sanjay Subrahmanyam, *The Career and Legend of Vasco Da Gama*, Cambridge: Cambridge University Press, 1998, p. 128.
3 This account along with many other subsequent descriptions of the mudbank can be found in File No. C-89, History of Mudbanks, Volume I, Cochin Government Press, Ernakulam, KSA—C, p. 68.
4 Ibid., p. 84.
5 See W. H. Roser, *The Seaman's Guide to the Navigation of the Indian Ocean and China Sea: Including a Description of the Wind, Storms, Tides, Currents, &c., Sailing Directions; a Full Account of All the Islands; With Notes on Making Passages During the Different Seasons*, London: Imray and Sons, 1867, p. 421.
6 On the enduring nature of this problem see Nancy Langston, *Where Land and Water Meet: A Western Landscape* Transformed, Seattle: University of Washington Press, 2003.
7 C-89, KSA—C.
8 Ibid.
9 Gotz Hoeppe, "What happened to the *chaakara*? The Formation of Coastal Mud Banks and the Reformation of Local Environmental Knowledge in Kerala (South India)," in Dilger, Hansjörg/Wolf, Angelika/Frömming, Urte Undine/Volker-Saad, Kerstin eds., *Moderne und postkoloniale Transformation: Ethnologische Schrift zum 60. Geburtstag von Ute Luig*, Berlin: Weißensee-Verlag, 2004, pp. 242–257.
10 See, for instance Hermann Gundert's *Malayalam-English Dictionary* published in 1872.
11 C-89, KSA—C.
12 Ibid., p. 73.

References

Carter, Paul. 1987 *The Road to Botany Bay: An Essay in Spatial History*. London: Faber and Faber.
Davies, Arthur. 1953 Loss of the Santa Maria, Christmas Day, 1492. *American Historical Review* 58: 854–865.

Dinesh Kumar, P. K. 2016 Mudbanks of Kerala: Mystery Yet to be Unveiled. *Science Reporter*, February: 38–41.

Hoeppe, Gotz. 2004 What Happened to the *Chaakara*? The Formation of Coastal Mud Banks and the Reformation of Local Environmental Knowledge in Kerala (South India). In *Moderne und Postkoloniale Transformation: Ethnologische Schrift zum 60*. Geburtstag von Ute Luig edited by Hansjörg Dilger, Angelika Wolf, Urte Undine Frömming and Kerstin Volker-Saad. Berlin: Weißensee-Verlag: 242–257.

Joseph, Mathew. 1992 Wave Mud Interaction in Mudbanks. PhD dissertation submitted to the Cochin University of Science and Technology.

Langston, Nancy. 2003 *Where Land and Water Meet: A Western Landscape Transformed*. Seattle: University of Washington Press.

Logan, William. 1989 *Malabar Manual: Volume I*. New Delhi: Asian Educational Services.

Mathew, K. J. and C. P. Gopinathan. 2000 The Study of Mudbanks of the Kerala Coast—A Retrospect. In *Marine Fisheries Research and Management*, edited by V. N. Pillai and K. G. Menon. Kochi: Central Marine Fisheries Research Institute.

Markley, Robert. 2007 Monsoon Cultures: Climate and Acculturation in Alexander Hamilton's a New Account of the East Indies. *New Literary History* 38(3): 527–550.

Menon, Padmanabha K. P. 1982 *History of Kerala, Vol.1*. New Delhi: Asian Education Services.

Nayar, Pramod K. 2005 Marvellous Excesses: English Travel Writing and India, 1608–1727. *Journal of British Studies* 44(2): 213–138.

Parizanganeh, Abdolhossein, V. C. Lakhan, and S. R. Ahmad. 2008 Dynamics of Mudbanks Along a Coast Experiencing Recurring Episodes of Erosion and Accretion. *Journal of Marine Engineering* 4(7): 70–79.

Robertson, George. 1875 On the Mud Banks of Narrakal and Allippey, Two Natural Harbours of Refuge on the Malabar Coast. *Proceedings of the Royal Society of Edinburgh* 8: 70–78.

Silas, E. G. 1984 *The Mudbanks of Kerala-Karnataka: The Need for an Integrated Study*. Cochin: Bulletin of Central Marine Fisheries Research Institute.

Subrahmanyam, Sanjay. 1998 *The Career and Legend of Vasco Da Gama*. Cambridge: Cambridge University Press.

3 "Source to Mouth"

Engineers, Rivers, Coasts and the Bengal Delta (1750–1918)

Rohan D'Souza

Introduction

"Change" was what preoccupied superintending engineer C. A. Williams in a compilation of notes that he published in 1919, titled as *History of the Rivers in the Gangetic Delta (1750–1918)*. These notes, as he averred in the preface, had been gathered from old records and by observations from tours that he personally carried out over a span of 13 years, a period when he held charge of the Northern Drainage and Embankment Division and the South-Western Circle in the irrigation department of the Bengal Presidency. This turn to history was not essentially an exercise in documentation. The chief object, Williams clarified, was to "forecast what is likely to happen in the future" and thereby be able to inform "future policy" (Williams 1919). The strong assumption here was that the historical record revealed regularities in the behaviour and moods of the delta's many raucous rivers. Prediction and control, hence, were possible. And therein lay Williams' hope for grasping the past through a future-orientation.

Tracking fluvial change, however, was wrapped up with another equally troubling challenge. The problem of coming to grips with environmental uncertainties. Ever since deltaic Bengal was brought under the full sway of the East India Company (EIC) by the latter half of the 18th century, the early Company officials had to hurriedly transform themselves from being an erstwhile trading concern dependent on mercantile profits to becoming instead a land-revenue extracting ruling authority. This shift, as Jon Wilson tells us, proved to be a fraught and troubled one. In particular, a slew of administrative problems immediately burst forward after the EIC instituted the *Permanent Settlement* of 1793: which by a single stroke of the pen legally constituted land as exclusive bourgeois property, a source for regular revenue and the basis for political stability that was to be realized in the person of the *Zamindar*—the

DOI: 10.4324/9781003282471-4

improving landlord. These former traders now turned into pioneering British colonial administrators, according to Jon Wilson, soon found themselves caught up in many a consuming struggle to establish "stable categories" that could help them make sense of the vast array of local "practices" involved in the collection of revenue (Wilson 2010, 104–132; Bhattacharya 2018). Not only did this, as yet incipient colonial bureaucracy, have to ascertain the revenue demand through newly instituted abstractions such as "permanent ownership" or "revenue in perpetuity" but, more pronouncedly, they came up against an unusually dynamic deltaic environment that proved to be economically and socially unsettling (D'Souza 2015, 2–24).

While the notion of land for Company officials was held to be enduring and constant, the rivers of Bengal's were considered the opposite: volatile and prone to extremes. Much to the surprise and bewilderment of many an EIC official, it was noted, for example, that even the most meagre of streams in the winter season could suddenly erupt during the summer monsoon rains into an enormous and torrential flow. The sprawling deltaic tracts, moreover, they soon enough realized, were incalculably cut through and run across by a wide spectrum of diverse hydraulic phenomena. The huge muscular rivers such as the *Ganges* and the *Padma* and *Meghna* not only threw voluminous offshoots like the *Hooghly* and the *Bhagirathi* but their massive flows also spawned a dense network of rivulets and minor channels (*khals*) that snaked and zigzagged across uneven ground before blithely emptying into an array of ponds and lakes. Towards the coasts, a vast collection of edge ecologies braided the interface between the fresh waters and the salty rhythmic tidal push that emanated from the Bay of Bengal. This biologically rich and bountiful interstitial zone was furthermore ringed with a complex and constantly rearranging maze of estuaries, creeks, marshes, swamps and tidal channels. And finally, as the waters would begin to empty into the Bay, they would first have to haphazardly weave across a huge labyrinthine sprawl of mangrove forests called the *Sundarbans*.[1]

The delta as a soil and liquid admixture, it increasingly became apparent, was a contingent soil–liquid arrangement that was seasonally recreated by floods, the violent shifting of river channels and the frequent and dramatic alteration of the boundaries between land and water by tidal actions and the non-too infrequent cyclonic storms. As the EIC matured into a full-fledged colonial bureaucratic dispensation a hard realization began to grow that the administration could give life to the *Permanent Settlement* only by splitting soils and liquids into two distinct non-overlapping domains and by seeing double: land as stability and rivers as uncertainty; land as the domain of political economy and rivers as the terrain of the technical; and lastly, land administered as property for revenue and rivers as resources to be harnessed through engineering.[2]

But much before the lens of revenue and property recast the rivers as temperamental and fickle forces, the latter was plotted as a fluvial transport highway. It was, in fact, the impatient quest in the early years for carrying out mercantile trade for merchant profit that drove the first big push for putting the rivers under the cartographic gaze that chiefly involved surveys and mapping exercises carried out by the daring of intrepid explorers.

Surveyor and Statistician

As early as 1764, the then Governor of Bengal Henry Vansittart (1759–1764) of the then fledgling EIC arrived at what seemed to be a very obvious realization: that if trade had to be secured in their newly acquired possessions within the Bengal delta, then the region's perplexing waterways needed to be reliably mapped and made carto-graphically coherent. A young and enthusiastic James Rennell (1742–1830)—and soon to become the first Surveyor-General of Bengal in 1767—was subsequently awarded a commission to not only survey the Ganges delta but, importantly, find a "shorter passage [that would be] suitable for large vessels from the Ganges to Calcutta" (La Touche 1910, 3). There were strong bets that Rennell's investigations would indeed end up sparing company vessels losses in time and money by discovering a short and safe route that could cut through the labyrinth of channels, rivulets and creeks in the Sundarbans. For EIC officials, who were then amidst many bitter battles to out muscle robust local trading networks, the Sundarbans loomed not merely as a physical barrier but as an enor-mously threatening living force.

Navigating the mangrove forests, islands and the mesh of waterways proved particularly baffling with many a dark surprise in that shipments were regularly lost to squalls, tides, pirates, shock currents and, often enough, by being mislaid by the unfathomable maze-like character of the journey. Almost a century later, the imperial chronicler Sir William Wilson Hunter (1841–1900) still felt compelled to admit that the Sundarbans continued to be unsurmountable

> as a tangled region of estuaries, rivers, and watercourses, enclosing a vast number of islands of various shapes and sizes…[and] nearer the sea, we find the primeval forest, impenetrable jungle, trees and

brushwood intertwined, and dangerous-looking creeks running into the darkness in all directions.

(Hunter 1875, 286–287)

But to return to Rennell's survey, one notes that besides his "frequent apprehensions" of tigers, he was, above all else, keen to underline that

> Next to earthquakes, perhaps the floods of the tropical rivers produce the *quickest alterations in the face of our globe* (italics mine). Extensive islands are formed in the channel of the Ganges, during a period far short of that of a man's life; so that the whole process lies within the compass of his observations.

(Rennell 1793, 335–364)

The rapidity at which such alterations happened is further corroborated in another noting in his journal, where he remarks with barely concealed bewilderment:

> The 25th [of September 1764]. at 4 PM came to Saatpour, y [the] Place where we left off the Survey of the River in June last. There had been so much of the Bank carried away by the Freshes (sic), that we hardly knew the place again; & could not have found the Mark out had it not been for a remarkable Tree which I formerly took ye bearings of.

(La Touche, 1910, 26)

And as to why the deltaic tracts were repeatedly churned and rearranged by fluvial currents, Rennell believed that two linked causes were principally responsible. First was the "irregularity of the ground" which made the rivers to "wander in quest of declivity" and second was the "looseness of the soil", which tended to easily "yield to the friction" or the erosive powers of the streams (Rennell 1793, 343–344). In effect, much of the soft alluvial deltaic tract was inherently unstable and prone to rapid shifts even by a single seemingly insignificant obstruction.

> Whether this be the fragments of the river bank; a large tree swept down from it; or a sunken boat; it is sufficient for a foundation: and a heap of sand is quickly collected below if. This accumulates amazingly fast: in the course of a few years it peeps above water, and having now usurped a considerable portion of the channel, the river borrows, on each side to supply the deficiency in its bed... Each periodical flood brings an addition of matter to this growing

island…and while the river is forming new islands in one part, it is sweeping away old ones in other parts.

(La Touche, 1910, 347)

That is, even a single minor obstruction on the bed of an insignificant channel within a season was enough to dramatically rearrange an entire swathe of topography.

His keenness for detail so endeared Rennell to the Company authorities that he was soon tasked with carrying out a full-scale cartographic exercise of the subcontinent, which in time resulted in the publication of the widely acknowledged and authoritative *Memoir of a Map of Hindustan* in 1783. For the historian Matthew Edney the *Map of Hindustan* quickly achieved a considerable intellectual standing, such that it replaced the earlier "multitude of political and cultural components of India with a single all-India state coincident with a cartographically defined geographical whole" (Edney 1997, 15). Put differently, though driven by the search for a short route to overcome what appeared to be a chaotic system of waterways, Rennell's maps actually ended up giving historical coherence to an idea of India as occurring within a singular geography.

It would be wrong to conclude, however, Edney went on to argue, that such mapping exercises from the late 18th century onwards were unconnected to the many political efforts to consolidate the colonial presence in the subcontinent. Rather, the tangible implications that followed from the innumerable cartographical exercises was about enabling the emerging British power in the subcontinent to become the "intellectual masters of the Indian landscape" (Edney 1997, 9–16). British India as a coherent geographical unit, in other words, was discursively assembled or conceptually constructed to refract colonial concerns about political power and economic dominance, rather than simply documenting geography and topography in a neutral manner. In effect, as Company officials energetically pursued cadastral surveys and mapped territories, identified resources, located villages, forts, roads and, most importantly, sited ports and marine landings along the chequered coasts of the subcontinent, their understandings about landscapes and people were invariably sought to be linked to efforts to lend coherence to colonial order and rule.

Ideas and notions about what characterized deltaic Bengal, hence, was as much a production of a particular colonial political imagination as it was a physical terrain that was fought over, mapped, surveyed and ruled. And it is by thus resituating the river systems, the alluvial plains and the different ecologies within political frameworks and ideological orientations that one can grasp how notions of change and uncertainty were woven into and placed within the colonial record. Notions about geography and the varied environments in colonial documentation, in other words, were not stable categories. The Sundarbans, hence, was observed in starkly different terms in the writings of James Rennell (first Surveyor General, 1767–1777) from that of William W. Hunter (first Director-General of Statistics, 1871–1881). Their perceptual differences marked as much by their differing political contexts as by the separation of time — a period of close to a century.

For Rennell, tasked as he was with finding navigable routes for Company shipping, the Sundarbans chiefly appeared as a morass of confusing waterways. And, in essence, the cause for riverine trade and merchant profit to becoming hostage to uncertainty, chance and fate. The only way out, at least in the estimate of the Company authorities, lay in finding an assured navigable route through a detailed and cartographically informed exploration of the mangroves. A full 100 years later, however, for Hunter, the delta's inherent instability was placed in an entirely different context. This time around, what was previously discussed as fluvial fickleness was juxtaposed against the British India government's then ongoing anxieties and quest for establishing a land-based regime for collecting revenue that was rested on a notion of permanent property. The historian Paul Greenough goes so far as to contend that a single essay of Hunter that was published in 1875 proved to have been singularly responsible for giving the entire Sundarbans bad press for decades to come. In a 60-page essay that was included in Hunter's first volume of *The Statistical Account of Bengal,* the Sundarbans was referred to as a "sort of drowned land, covered with jungle, smitten by malaria, and infested by wild beasts" (Greenough 1998, 240). The implication was that as a landscape the Sundarbans was, conceptually speaking, a "wetland frontier" that would be hostile to settled agriculture, to the possibilities for realizing revenue and environmentally inimical to advancing colonial social institutions and the market economy (263).

In other writings on the Bengal region, however, Hunter put forward a much more process oriented explanation for the deltaic tract's inherent instability. The Ganges, he wrote, when it courses through the soft alluvial plains of Lower Bengal is on the "third stage of its life", having earlier traversed through the broad plains of Northern India and upon being further charged by many of her tributaries becomes the "unwearied water-carrier" for the people settled alongside its banks (Hunter 1875, 51). But as soon as the silt and detritus laden Ganges rumbles onto the levelled plains of Lower Bengal, the speed of its flow is greatly diminished and its channel begins to choke with the deposition of its burden onto its own bed. The Ganges then, Hunter notes, "splits out into channels, like a jet of water suddenly obstructed by the finger, or a jar of liquid dashed on the ground" (55). Unlike the flood plain, when tributaries pour into the Ganges in the deltaic stage, the entire fluvial regime is suddenly turned on its head with the deteriorated main stem now itself imploding and shooting-off arms or distributaries, which further splinter into innumerable streams, minor channels and creeks that sluggishly wind their way into stagnant lakes, swamps, marshes and estuarine bodies before scattering into the Bay of Bengal. At heart, however, what was actually unravelling was a geological process, wherein Lower Bengal, as Hunter surmised, receives

> each summer a top-dressing of new soil, carried free of cost for more than a thousand miles by the river currents from Northern India or the still more distant Himalayas—a system of natural manuring which yields a constant succession of rich crops.
>
> (Hunter 1875, 56)

Put differently, the silt-laden flows were dynamic geologic agents that ceaselessly and tirelessly were building-up deltaic Bengal through inexorable processes of erosion and deposition.

Once grasped as geologic process, however, the rapid topographical change and fluvial uncertainty within the Bengal delta appear as means towards realizing a purposeful geological end. That is, environmental dynamism and instability in the delta—brought on by river action, floods and the constant shifts between land and water—could be described as being part of a natural and necessary geological momentum that ultimately leads to land formation. By the early decades of the 19th century, the field of geology had begun to evoke considerable interest and received wide intellectual acceptance in Europe. The Scottish lawyer Charles Lyell (1797–1875), following the publication of his three volume *Principles of Geology* (1830–1833), had made the study of a changing earth an exclusive endeavour of science and reason. Central to which was the elaboration of what was referred to as the uniformitarian position. That is, the belief that causes that operated in the past in changing the earth continued to shape processes well into the present. Accordingly, the earth's geological changes, it was held, drew principally from two causes: *aqueous* (erosive actions of rivers, torrents, streams, currents and tides) and *igneous* (renewal by volcanoes and earthquakes).[3] The play of both causes in tandem, Lyell believed, operated to create a continuous balance—a mix of destruction and replenishment—that ended up producing an approximate constancy of sorts. The uniformitarians, moreover, saw themselves as challenging the claims of catastrophism. A leading proponent of which was the French geologist and naturalist George Cuvier (1769–1832), who maintained that the earth's geological features, unlike the gradualism of the uniformitarians, were a result of sudden breaks and oftentimes by violent ruptures (Palmer 2003, 23–35).

While Hunter's discussion of the rivers, the alluvial tracts and general assessments about land formation in Bengal suggests that notions of gradualism and uniformitarianism had most certainly informed his views, as might have equally been the case with others in the colonial government. On the other hands, the frequent shocks delivered by floods, sudden events of inundation and the rapid shifting of river channels, would have also, in all likelihood, garnered a number of adherents for catastrophism. What, nonetheless, appears clearer is that through the course of the 19th century the colonial administration steadily began to confirm that they were sitting atop, if not contending against, an enormous and active geologic process in the Bengal delta rather than merely dealing with an erratic hydrology or a static landscape that suffered from poor drainage brought on by a chaotic river system.

Change Leads to Centralization

Superintending engineer C. A. Williams' notion of history, consequently, had to above all else conceptually come to grips with the delta's muscular and process-driven environmental dynamism. The narrative essentially had to insert geography as deep-time-geological process within a meaningful unit of human-scaled historical time, even as the study was to be kept steadfast and focussed on tracking the ecological and political impacts from infrastructural interventions by the colonial government. *History of the Rivers,* thus, took a less than modest slice of 150 years to focus on a section that roughly comprised 19,000 square miles of the Bengal delta—"South of the Ganges down to the sea face... [that was] ... bounded on the west by the Hooghly and the Mathabhanga and on the east by the Padda [*Padma*] and Megna [*Meghna*]" (Williams 1919, 1).

Though engineer William's understanding of geological process via historical narrative was simple minded, his efforts could be uncannily connected to the later scholarship of Fernand Braudel (1902–1985), the celebrated French historian and one of the leading lights of the *Annales School.* Braudel elaborated a tripartite notion of historical time, which he first briefly spelled out in the preface of his magisterial *La Méditerranée et le Monde Méditerranéen à l'Epoque de Philippe II,* published in 1949. Accordingly, historical time was differentiated into "successive levels": the "surface disturbance" of the political event or individual time; the conjuncture of recurring cultural rhythms and economic cycles or social time; and finally, the *longue durée* or geographical time that refers to the "almost changeless" involving "man in relation to his surrounding" and his "contact with the inanimate" (Braudel 1982, 3–4). In a later essay published in 1958, Braudel, however, in a "call to discussion" with the social sciences sought to rethink the notion of the *longue durée* by now choosing to recast them as "hindrances", "limits" and as kinds of "structure" that implied "...difficulties of breaking out of certain geographical frameworks, certain biological realities, certain limits of productivity, even particular spiritual constraints: mental frameworks too can form prisons of the *longue durée*" (Braudel 1982, 31).[4]

But in Williams' narration, geography in the Bengal delta was neither a limit, a hindrance, nor a ponderous slow moving or near-frozen environmental backdrop, standing almost outside of historical time. Instead, the delta's soft alluvial plains with its raging shape-shifting rivers fitted more aptly into Braudel's notion of the political event—the "short, sharp, nervous vibrations" in which the "slightest movement sets all its gauges quivering" (Braudel 1982, 3). The near instant ecological impacts, in fact, is what Williams assiduously details when he observes the geological process of land-building being interrupted by the construction of physical infrastructures. The building of roads and railways not only "caused the death of many streams", he records, but the frequent "cuts" made across the "loops of rivers" also resulted in the sudden drying up of the surrounding streams and rivulets.

But amongst all the infrastructures that proved to be most vexatious in the delta were the flood control embankments. These embankments, Williams lamented, "raised problems as great if not greater" than what they were expected to solve. These structures not only aggravated flood levels by preventing the silting or raising up of the surrounding lands but also caused the rivers to choke their own beds with the silt and debris that they prevented the rivers from transferring out of their channels. Besides, many of the embankments were poorly situated and ended up interrupting local drainage and causing "waterlogging and severe epidemics of malaria and other diseases" (Williams 1919, 85–86). In sum, by artificially forcing and sustaining the divide between land and water, floods in the Bengal delta were being aggravated and becoming the main source for continual grief and suffering of the local populace.

Having thus surveyed environmental changes, ecological impacts and infrastructural interventions within a single historical frame , Williams could then firm up on a single conclusion: "rivers cannot be dealt with satisfactorily piecemeal" and therefore the "whole system from the source to the mouth" had to be brought under "one central controlling authority" and to whom all "questions should be referred which relate to any works affecting any river" (Williams 1919, 87). In effect, managing the Bengal delta could no longer be allowed to depend on the chaos of local arrangements and interests but needed a top-down expert-led centralized authority that could command flows at the level of the entire delta. The deltaic plains, the rivers and the coasts, therefore, had to be grasped as a single interconnected unit that, in turn, called for an organizational gaze from above.

Conceptualizing the Bengal delta as an environmental whole made up of interlocking ecologies, interestingly enough, began to garner appeal throughout the early decades of the 20th century, especially amongst engineers. To mention only one, S.C. Mazumdar (1884–1956), who served as a chief engineer of Bengal in the 1940s, in a lecture titled *Rivers of the Bengal Delta* delivered at Calcutta University and published in 1941, further extended the ecological boundaries of the delta by drawing connections between the rivers and their catchments in the hills. The notion of the "source to mouth", in effect, for Mazumdar, was evident in the fact that the Bengal coasts were now connected even further upwards towards the hilly catchments from where the waters of the rivers actually originated (Majumdar 1941, 2–29). The policy implication, for Mazumdar was the seemingly now obvious argument for an "Inter-Provincial Commission", which, he believed, could control the catchment basins and develop plans to overcome "isolated actions" (23). That is, the Bengal delta was now scaled to the level of the region—compacted as an agglomeration of hills, rivers, the alluvial plains and coasts—that earnestly called out for a more comprehensive, expert-driven and planned organizational intervention.

In conclusion, though this essay is a somewhat perfunctory discussion of a few colonial surveys and assessments of the Bengal delta, it does, nonetheless, helps us understand how a fascinating struggle erupted over the need for a meaningful environmental scale for intervention. Engineer Williams' turn to history, I suggest, enabled him to argue that only a centralized expert-driven organization could mitigate the effects of geological uncertainty, recurrent landscape change and the loss of predictability. Consequently, managing environmental dynamism, for him, could not be left to local decision-making and interests who not only haphazardly carried out embankment construction but had been equally unthinking in how they built roads and railway lines that ended up interfering with the flow regimes of rivers.

The early decades of the 20th century increasingly consolidated a strong engineering vision that viewed the delta as a huge complex of interconnected ecologies, one that wove the hills, rivers, alluvial plains and the coasts into a single natural unit. The technical upshot of this was that only upon conceptualizing the delta at such a massive environmental scale could a compelling case actually be made for controlling and harnessing the rivers system through centralized engineering organizations and mammoth infrastructures. The steady consolidation of this large-scale view of deltas, in fact, proved especially critical from the late 1940s onwards. A period which witnessed the aggressive introduction of the Tennessee Valley Association's inspired multi-purpose river valley development projects in Eastern India (Ekbladh 2010): notably, the construction of a string of large dams under the aegis of the Damodar Valley Corporation and the Hirakud Dam.[5]

Notes

1 For an excellent discussion of the geology and hydrology of the Bengal delta, richly supported with both early and contemporary maps, see Kalyan Rudra, *Atlas of Changing River Courses in West Bengal*, Kolkata: Sea Explorer's Institute, 2012. Also see Iftekhar Iqbal, *The Bengal Delta: Ecology, State and Social Change, 1840–1943*, Palgrave Macmillan, 2010.

2 On what deltaic flooding meant to colonial revenue policies and administrative constraints see Rohan D'Souza, "Event, Process and Pulse: Resituating Floods in Environmental Histories of South Asia", in *Environment and History* (Special Issue: Disasters and the Making of Asian History edited by Chris Courtney and Fiona Williamson), 26, 2020, pp. 31–49. On urban flooding in the colonial city of Calcutta see Debjani Bhattacharyya, *Empire and Ecology in the Bengal Delta: The Making of Calcutta,* Cambridge: Cambridge University Press, 2018.

3 On this see Stephen Jay Gould, *Time's Arrow, Time's Cycle: Myth and Metaphor in the Discovery of Geological Time*, Penguin Books, [1987] 1991, p. 146.

4 This essay was republished from Fernand Braudel 'History and the Social Sciences: The *Longue Durée'*, *Annales* 4, October–December 1958, pp. 725–753. For a lively discussion on the role of the *longue durée* and contemporary politics see David Armitage, and Jo Guldi, 'For an "Ambitious History" A Reply to Our Critics' in *Annales. Histoire, Sciences Sociales* 70(2), 2015, pp. 367–378. Also see Jo Guldi and David Armitage, *The History Manifesto*, Cambridge University Press, 2014.

5 On the Hirakud dam see Rohan D'Souza, *Drowned and Dammed: Colonial Capitalism and Flood Control in Eastern India,* New Delhi: Oxford University Press, 2006, and for the DVC see Daniel Klingensmith, *'One Valley and a Thousand': Dams, Nationalism and Development*, New Delhi: Oxford University Press, 2007.

References

Armitage, David, and Jo Guldi. 2014 *The History Manifesto*, Cambridge: Cambridge University Press.

Armitage, David, and Jo Guldi. 2015 For an "Ambitious History": A Reply to Our Critics. *Annales Histoire, Sciences Sociales* 70(2): 367–378.

Bhattacharya, Neeladri. 2018 *The Great Agrarian Conquest: The Colonial Reshaping of a Rural World*, Ranikhet: Permanent Black.

Bhattacharyya, Debjani. 2018 *Empire and Ecology in the Bengal Delta: The Making of* Calcutta, Cambridge: Cambridge University Press.

Braudel, Fernand. 1982 *On History*. Translated by Sarah Matthews [1980]. Chicago: University of Chicago Press.

D'Souza, Rohan. 2006 *Drowned and Dammed: Colonial Capitalism and Flood Control in Eastern India*, New Delhi: Oxford University Press.

D'Souza, Rohan. 2015 Drainage, River Erosion and Chaurs: An Environmental History of Land in Colonial Eastern India, *Nehru Memorial Museum and Library Working Paper Series: History and Society*, New Series 84: 2–24.

D'Souza, Rohan. 2020 Event, Process and Pulse: Resituating Floods in Environmental Histories of South Asia. *Environment and History* (special issue: 'Disasters and the Making of Asian History,' edited by Chris Courtney and Fiona Williamson), 26: 31–49.

Edney, Matthew. 1997 *Mapping an Empire: The Geographical Construction of British India (1765-1843)*. Chicago and London: University of Chicago Press.

Ekbladh, David. 2010 *The Great American Mission: Modernization and the Construction of American World Order*. Princeton and Oxford: Princeton University Press.

Gould, Stephen Jay. 1991 *Time's Arrow, Time's Cycle: Myth and Metaphor in the Discovery of Geological Time*. London: Penguin.

Greenough, Paul. 1998 'Hunter's Drowned Land: An Environmental Fantasy of the Victorian Sunderbans'. In *Nature and the Orient: The Environmental History of South and Southeast Asia*, edited by Richard H. Grove, Vinita Damodaran and Satpal Sangwan. New Delhi: Oxford University Press.

Hunter, William Wilson. 1875 *A Statistical Account of Bengal: Volume I*, London: Trubner & Co.

Hunter, William Wilson. 1882 *The Indian Empire: Its People, History and Products*. Revised 3rd edition. London: Smith Elder & Co.

Iqbal, Iftekhar. 2010 *The Bengal Delta: Ecology, State and Social Change, 1840–1943*, Basingstoke: Palgrave Macmillan.

Klingensmith, Daniel. 2007 *'One Valley and a Thousand': Dams, Nationalism and Development*, New Delhi: Oxford University Press.

La Touche, T. H. D., ed. 1910 *The Journals of Major James Rennell: Written for the Information of the Governors of Bengal During His Surveys of the Ganges and Brahmaputra Rivers, 1764-1767*, Calcutta: Baptist Mission Press.

Majumdar, S. C. 1941 *Rivers of the Bengal Delta*, Calcutta: Calcutta University Press.

Palmer, Trevor. 2003 *Perilous Planet Earth: Catastrophes and Catastrophism Through the Ages*, New York: Cambridge University Press.

Rennell, James. 1793 An Account of the Ganges and Burrampooter Rivers. In *Memoir of a Map of Hindoostan: On the Mogul Empire*. 3rd edition. London: W. Hulme: 335–364.

Rudra, Kalyan. 2012 *Atlas of Changing River Courses in West Bengal*, Kolkata: Sea Explorer's Institute.

Williams, Addams C. 1919 *History of the Rivers in the Gangetic Delta 1750-1918*, Calcutta: Bengal Secretariat Press.

Wilson, Jon E. 2010 *The Domination of Strangers: Modern Governance in Eastern India, 1780-1835*, London: Palgrave Macmillan.

4   Living Paradox in Riverine Bangladesh

Whiteheadian Perspectives on Ganga Devi and Khwaja Khijir

Naveeda Khan

Whitehead within the Riverine Context

In a village in the southeastern part of Bangladesh, a woman tells a researcher about her state of mind at the time of her son's death by drowning in a pond close to their homestead:

> I thought about nothing, had no memory of anything. Now I know (boy's name) is not alive. That day I cooked with such attentiveness it was as though I was childless. After finishing, I remembered that I hadn't seen (child's name). The other day he came to me many times, he touched the wood and searched for fish to eat. That day he did not come and I did not even notice.
>
> (Blum et al. 2009, 1723)

This research article describes how respondents subscribe to the idea that supernatural beings live in the water and seek continual sacrifices or appeasements, sending out enticements to people. Says another woman of her young daughter's drowning death:

> She didn't die due to drowning. Evil took her to the water and killed her. If anyone digs a new pond they must give some rice, egg, turmeric, chira (rice concoction), nose ornaments, and coins to the pond. But we did not do it so Gongima became dissatisfied and wanted a human being in revenge.
>
> (Blum et al. 2009, 1723)

And it is not only women in the southeastern part of Bangladesh who speak thus. I also heard such charges against Ganga Devi, as Gongima is also called, in the northern part of Bangladesh where I researched on the sand and silt bars called *chars* that regularly form and erode in

DOI: 10.4324/9781003282471-5

the middle of the Jamuna River, the largest of the three major rivers in Bangladesh. And here she was not alone in bearing the blame. Her consort, Khwaja Khijir, also known as Khidr, was similarly considered to entice children to early watery deaths.

Bangladesh is on the largest delta of the world, the Ganges Delta, which empties into the Bay of Bengal. It is crisscrossed by rivers, ponds, ditches, embankments, and lakes and faces monsoon rains, floods, and cyclones. It is one of the most watery landscapes on earth with 7% of its surface under water all the time and two-thirds under water some of the time. Consequently, most people live in some proximity to water, and, of course, those who live on *chars* live, by definition, on the water. This landscape is the very definition of Terra Aqua. Drowning deaths are prevalent, particularly among the young between the ages of one and five. Such deaths are largely attributed to the inadequate supervision of children. Efforts are underway to have creches or even cribs available for young mothers, who bear most of the burden of housework in rural homesteads, rich and poor alike (Blum et al. 2009).

Despite the seemingly self-evident explanation for children's deaths by drowning—that is, ready access to standing water and inadequate supervision—two aspects of the women's responses are surprising and worth reflecting upon. First, there is the insistence by women that they had fallen into a state of forgetfulness, even indifference, while their children were lured to their watery deaths. This is quite unlike the case of the shantytowns in Brazil in which Nancy Scheper-Hughes (1992) has shown how the high rates of child mortality were met by a cultivated indifference by mothers toward their deceased babies. For one, child mortality has reduced substantially in Bangladesh in the past two decades. Furthermore, maternal love and care for the child by caring for oneself and the baby is encouraged from the commencement of pregnancy (Blanchet 1984). While mothers may not be able to lavish attention on their children, there is in fact a cultural inhibition against showing such open and preferential treatment, mothers are always mindful of their children's needs (see also Trawick 1992).

In addition, the death of each child remains with the mother. While child mortality rates in the *chars* is undoubtedly higher than in the adjoining areas on account of the lack of medical services, the difficulty of movement because of non-existent infrastructure and the general poverty of the population, *chauras*, or those who live on the *chars* (sandbanks), mourn the loss of their children, even the ones they were not able to know or raise, marking such deaths with *milads* or prayer gatherings that they can ill afford. So, the idea that mothers fall into a state of indifference signals less a state of their distraction wrought by busy lives and more the fact that they were compelled to forget and that they were controlled by some presence. Their insistence endures in the face of criticism by their husbands and in-laws, and it is set apart from their understanding of the deaths of children by other means. For instance, if a child does not survive infancy, they are as likely to say that the child lacked an attachment to life as the mother did not eat nutritiously when pregnant or did not feed the child nutritious food. Thus, the insistence that the children were lured by Ganga Devi, Khidr (*Khijir*), or other figures to their drowning death seems to comprise a class of explanations onto itself, expressing the idea that such a death is not inevitable or accidental—it does not arise from within the child or due to the mother's irresponsibility—but, rather, that it is brought about through the action of external forces. Therefore, at least in this class of explanations, there is a very strong sense of being affected, not by God, the all-transcendent figure who acts upon all of humanity, but, rather, by forces only prevalent here and strongly associated with water.

The second noteworthy aspect of the women's explanations is that they blame Gongima, a Hindu deity, for the loss of their children. The surprise in this lies in the fact that the women are all Muslim and have had little or no contact with Hindu traditions since many Hindus left this region for India or elsewhere during the 1970s due to the persecution they encountered when Bangladesh was still a part of Pakistan (Jahan 1972). Others simply left because their villages eroded as the Jamuna River moved westward, and they never returned after their lands came back. Ganga Devi, as Gongima is also called, has a long history in Bengal as both a deity and, in her manifestation on earth, as a natural entity—a river—yet one is hard pressed to find any temples, rituals, or invocations of her in present-day Bangladesh (Darian 1978).

It is not as surprising that the women also blame Khwaja Khijir, also referred to as Khidr, because he is associated with the Islamic tradition and has a presence in other Muslim riverine communities, but the women's charge against him is still surprising because he has traditionally been seen to be a protective figure, not one who deals in death (Albinia 2010). Also surprising is that he is Ganga's consort, suggesting the need for some explanation as to how the two come to reach across two different traditions to be together. Syncretism, a common explanation (Roy 1984), is not entertained here because, as Richard Eaton (1996) and Tony Stewart (2001) have shown empirically in Bengal, Hinduism and Islam have never been combined to produce a third way.

Instead, conversion has happened through the seeking of equivalence between figures across the two traditions, and then, either the steady eclipse of one by the other or the insertion and encasement of entirely different traditions within another. Carlo Severi (2004) writing on the Apache ghost dance in North America has further shown how two traditions may exist in parallel, without syncretism or even conversion, through a relationship of paradox. For example, when the Shaman-Messiah, who had come to lead the Apache ghost dance, claimed himself to be Jesus; he was not inserting himself into the religion of the Whites or producing a new syncretic form of Jesus or putting himself in a relationship of equivalence to Jesus, but, rather, he was saying: "If I am similar to you, then I am different." What makes this not simply a contradiction, but a paradoxical statement is that "a logical link is established between two contradicting predicates" (820). Thus, for Ganga and Khidr to be invoked together in the context of drowning deaths requires some inquiry into the paradoxical relations between the two and the possibility that paradox emerges from the river waters.

How then do we understand this unusual situation in which women explain their forgetfulness and the deaths of their children as being caused by a goddess who is no longer animated in their social milieu, one who has no referent within their usual frameworks for understanding their lives, and by a mythic prophetic figure who is also on the wane within the local practices of Islam? The research study on child mortality and other such studies imply that this mode of explanation is a survival of the past, one steeped in superstition, and that given sufficient education on the rightful causes of drowning deaths, these women will be able to overcome the hold of this mode of thinking. In this article, I ask instead that we entertain the possibility that the women's expressions help us to understand the present as a palimpsest of prior forms of life, each a fragmentary reference opening into a stratum of past modes of being and their permutations. I ask also that these fragments not be treated as survivals of a past in which Ganga and Khidr were lodged in people's lives and imaginations but, rather, that the continued, now much-enfeebled, reference to them suggests a standing means by which to produce bridges to the future and to discern patterns within the swirl of uncertainty that is the future. Given that Ganga and Khidr are so strongly associated with water, specifically that of the river, I suggest that we stand to not only understand women's changing relationship to the river through following their relationship to Ganga and Khidr and the relationship between Ganga and Khidr but that we also stand to learn something of the river's own changing nature. The river's physical transformation is laid bare for us along this stratum of changes and exchanges.

When I suggest that we can see the river through the women's discourse, I am not saying that the river is available to our analysis through its representation such discourse but, instead, that their discourse is waterlogged, suggesting their embeddedness in watery environments. How are we able to claim such an immediacy of expressivity without the mediation of cultural frameworks and mental categories that have come to be established as the sine qua non of representation? Alfred North Whitehead, the scientist and philosopher noted for his commitment to propounding a philosophy of process, is salient here in allowing us to get beyond the dualism of representation and unmediated immediacy to consider how cultural categories—or what he calls symbolism— and physical processes work together. It is important to sketch some basic tenets of his thought to show how he achieves an immediacy between culture and nature rather than the outmoded notions of environmental determinism—the presumption of the historical domination of culture over nature or the comingling of the two as natureculture that is the standing position within anthropology.

Here I follow Whitehead's (1979 [1929]) ontological principle articulated in Process and Reality that

> "actual entities"—also termed "actual occasions"—are the final real things of which the world is made up. There is no going behind actual entities to find something more real ... And these entities are drops of experience, complex and interdependent.
>
> (18)

When these women speak in this way, they speak metaphysically (metaphysical is meant here in the sense of the supernatural), and based on the ontological principle, any metaphysical explanation must incorporate actual entities and occasions or else they would be deemed inadequate and revised (13). In other words, the women's descriptions must attend to their own experiences and those they have had before them that are congealed within other actual entities and occasions that they make their own. Within Whitehead's framework, it is through experience that existence unfolds in the world, with the women experiencing the lure of propositions from the world, undergoing a series of feelings called prehensions, to arrive at a feeling of self-satisfaction or concrescence. This process is called an actual occasion with the women being a composite of many such occasions, including the occasion of the deaths of their children and that of narrating their deaths.[1]

This is a very schematic snapshot of Whitehead, but even in its "basicness" it is interesting placed alongside the recent ontological turn within anthropology (Kohn 2015). Whitehead would seem to support the idea of multiple worlds—in this case, of the women and children from the men and the elderly (see de Pina-Cabral 2014a, 2014b; Strathern 2006), in so far as the women's prehensions align them into a nexus or society among themselves to the exclusion of those who do not have similar prehensions (Halewood 2014). However, the purity of separateness is not tenable within Whitehead as the women only seem to form a society with other women along certain lines of experience and expressions, while being in societies with children, men, and the elderly along other lines and, as I hope to show, in societies with prophets and goddesses and with silt and water. Whitehead thus presents a picture of all of us as participants of many societies at once but within one world.[2] Note also that Whitehead is not speaking of humans or beings in their full consciousness, but, rather, he is attempting a descriptive apparatus that may be equally applied to a squirrel or an electron.

To return to my initial provocation, when the women speak about the presence of supernatural beings and their hold upon women and children, they are expressing their watery environment. However, Whitehead cautions, while we may begin with their verbal expression, we must beware that language is problematic in that words are vague and ill-defined. We cannot go from their expression to how the world draws the women in. I read Whitehead as suggesting that while verbal expressions comprise both objective content and subjective form, those of religious or metaphysical nature such as the women's demand closer attention to the subjective form or the emotive feelings they express and arouse in the speaker and the hearer. What this means in a Whiteheadian framework is that the women's verbal expressions speak more to enduring eternal objects—another order of things in addition to actual objects or occasions—than to their immediate reality, but which are also active within the women's unfolding existence and actual occasions.

The feelings associated with eternal objects are more conceptual than physical feelings, suggesting that the women's relationship to the riverine environment, at least in this instance, is more determined by history, tradition, religion, and culture than with their physical responses to it. Thus, far from being an error in judgment, the emplacement of the supernatural being at the site of their children's watery deaths speaks to the women drawing upon mythology as a long-standing resource. It is therefore this closeness of eternal objects with actual occasions that makes it possible for us to move in this article from the women's verbal expressions of their mythological endurances to the actual occasions of the feelings and experiences of the riverine environment, toward the particular occasion of the river as a physical entity, and back again.

I find this approach to the women's expressions to be more productive than other approaches within environmental studies and anthropology, in which women's expressions may be read as either risk perception (Amoore 2013; Douglas and Wildavsky 1983) or danger behavior (Steiner 2013 [1956]). The modern language of risk brings in a presumption of calculative logics at work within the women's perceptions, and the older language of danger behavior assumes clearly articulated taboos. These are overly deterministic, whereas expressions and experiences may be more inchoate and indeterminate. While the supernatural being may be a figuration of fear and dread, to focus on fear alone is to reduce the range of emotions, states, and actions to which the figure also provides access.

While I pursue folklore and mythology in this paper by studying the changing permutations of the figure(s) of the supernatural, I do so somewhat differently than the usual studies in which folklore and myths are interesting only in themselves and in what they have to say about the structure of thought or the evolution of ideas, thus putting a wedge between being and the world.[3] A Whiteheadian informed perspective is careful not to perpetuate the fallacy of thinking that we are speaking of a sole tradition, a corporeal experience, a single event, or a simple location when speaking of actual occasions and eternal objects. It insists upon maintaining a continuous loop of interpretation through being and the world.

In what follows, I track the various permutations of supernatural beings introduced to us by the women, specifically the figures of Ganga and Khidr, across space and time, to show how it is paradox, other than risk, danger, or fear, that comes to be privileged, experienced, and inflected within the watery environments and riverine lives in the Jamuna. Paradox is, to remind the reader of Carlo Severi's (2004, 820) words, "a logical link . . . established between two contradicting predicates." However, here paradox is situated in the world, as a lure from the world, rather than housed within individual thought or psyche. To feel and remark on paradox is to note nature's working through one. This feeling—the tug of paradox—is productive in so far as it suggests both an openness to one's environment as well as the internal recursion and bifurcation of paradox within selves and societies. In other words, the feeling of paradox makes manifest a realm of experiences of the river that we cannot get at otherwise and that constitutes a part of the story of water that we need to tell in the face of its ongoing abstraction as a mere resource (Hastrup and Rubow 2014). Thus, an important claim of this paper is that paradox serves as the occasion for the togetherness of the mythological, the physical, and the social. Among some of the paradoxes we encounter as ordinary and everyday are those that sustain the relationship of Ganga and Khidr and that permute through Khidr but that also speak to people's experiences of the river and the truths of the river itself—such as, if one is dead one is immortal; if one is separate from another, one is immersed in the other; in cruelty there is mercy, in stillness there is movement—until we arrive at what I feel to be the portentous element within the women's expressions that speaks to the possible future of the river within the context of the changing climate, notably in its forgetting, where there is remembering.

Ganga Ma and Khwaja Khijir: The Event of Paradox

As noted earlier, the supernatural being that the women in southeastern Bangladesh refer to as Gongima is also called Ganga Ma or Ganga Devi in other parts of Bangladesh, the appellation for the personification of the sacred River Ganges. She is often pictured as a beautiful woman sitting astride a crocodile or fish or an amalgamation of both known as Makara.[4] The most persistent representation of her that I came across was with Khwaja Khijir or Khidr, also mentioned earlier, within the context of a festival called Bera Bhasan. Khidr is often pictured as a white-bearded man standing astride a fish. The twinning of the two provides insight into the way in which Islam found footing alongside Hinduism in Bengal.

During one of my early stays on a *char* within the Jamuna River, where I carried out my work, I accompanied a young male employee of the non-governmental organization (NGO) that was hosting me on his weekly visit with one of his groups. These *chars* are geophysically made up of the sedimentation brought down by the river waters. The sediment is deposited along various filaments of the braided river, either alongside existing land or in island formations, and can exist anywhere between 3 and 15 years on average (Environment and Geographical Information Systems 2000; Islam 2010). Their relative temporariness is provided as the explanation for why *chars* are not well serviced by the Bangladeshi government. There is almost no infrastructure such as roads, bridges, water supply, and electricity grids. The people who live on them are either among the landless poor who cannot afford to live on the mainland, or those who claim that the *chars* house their villages and properties returned from the river. While farming and fishing are the main occupations on the *chars*, men living on them must seek work in other industries (such as brick making, weaving, construction) outside of these areas to sustain an annual living. Women, if they work outside of their homesteads, may take to cutting roads or working in Bangladesh's extensive garment industry. Otherwise, they are busy at home collecting water, cooking food, washing clothes and dishes, sweeping the household, and tending to their children, gardens, and animals. While NGOs precisely target this income group, known as the "hardcore poor," and provide them with microcredit loans and other provisions, the larger, more mainstream NGOs are leery of loaning to *chauras* who are considered to be at a risk for flight as they are prone to moving because of the vulnerability of their homes and lands to erosion and floods. This vulnerability is often read back upon their character, marking them as untrustworthy and immoral. However, I noted a much stronger streak of social conservatism within *chars* than in the villages on the mainland in that women have a very limited sphere of movement. Even if they go to cut earth, they travel in groups, and when they return, they again restrict themselves to their households and immediate neighbors. At the same time, there is a more relaxed attitude toward illegal activities that are marked as immoral, such as drug use and gambling by the men and even occasional acts of violence. This combination of social conservatism and the acceptance of immoral behavior is an indication of the ordinary ways in which gendered contradictions are manifest and managed within *chaura* lives.

The NGO hosting me was very local, created by those who are from *char* backgrounds, who know well the local histories and networks of people, and who can track down debtors through these links. Furthermore, they have in place mandatory savings for their members that are treated as collateral against loans. My guide, the young man from the NGO, was there to collect weekly payments on microcredit loans made by his NGO to these group members as well as their savings. After completing these duties, he carried out a short workshop on disaster preparedness, a central mode by which to render the *chaura* population more resilient to their environment but one with ambiguous effects (Khan 2014).

This group was well suited to my interest in studying how people cope with erosion as they had recently lost a large part of their village to the ravages of the river and were expecting to lose all of it soon, leaving them no choice but to move elsewhere. I found the group to be comprised of tightly knit family members who were proud of their *bongsho* (patrilineality) and village (patrilocality). Although they were moving less than half a mile away to a place we could see with our naked eyes, they considered themselves to be moving to *bidesh* or foreign lands, suggesting how tightly drawn were the lines of belonging and order, with disorder prevailing outside of them.

In the context of being asked what they do to prepare for erosion and other natural disasters upon their lives, the women faithfully repeated the words of advice that they had just received from the young NGO worker. They were expected to have set aside money, food staples, and a portable mud stove should they need to move. At the same time, they recalled a childhood festival of the Bera (raft) Bhashan (sending out), in which in a day in *pous* (January), after the monsoon and harvest seasons, they would float large leaves of the banana tree bedecked with flowers and sweets to expiate Ganga Devi and Khwaja Khijir. They insisted that it was always done for the two of them and not for one alone. They giggled, recalling the silly songs they sang about the upcoming nuptials as they pushed off their makeshift rafts:

> *Dol dol doloni*
> *Ranga Mathayey Tchironi*
> *Bor Ashbey Akhoni*
> *Niye Jabey Tokhoni*

> *Swing Swing Swing*
> *A comb through your bright head*
> *Your groom is coming at any moment*
> *He will take you right away*

It seemed that these festivals were no longer ongoing in the silt island, although the festival of Bera Bhashan is still held in different parts of northeastern Bangladesh, where elaborate rafts are constructed with the trunks of banana trees, richly decorated, and set afloat amid much singing and fanfare (Sayidur 1991). However, these are much reduced in pageantry from sixteenth-century Mughal India, which saw an efflorescence of the festival as the Arab Muslim mythical figure of Khidr (the Green One), first introduced into Bengal in the thirteenth century, came to hold his own against the pantheon of Hindu gods and goddesses, of whom Ganga Devi was one (Haque 1995 [1975]). In the Mughal period, the festival was less about the expiation of Ganga or Khwaja (Lord), as Khidr came to be called in Bengal, and more a marking of the confluence of maritime and agrarian lives and the temporalities of trade, fishing, and harvest through an invocation of Khidr as the protector of all those who plied the waters (Mukherjee 2008).

Khidr had come to this appellation through an earlier series of associations with Alexander the Great's cook, who had imbibed the water of life rather than give it to Alexander, thereby gaining immortality; and within the Christian tradition, in the stories of the Prophet Elias or Elijah (Longworth Dames 2013). Within India, he was the equivalent of the Vedic god Varuna, the god of the waters and the lawgiver to the underwater world (Haque 1995 [1975]). While it was Khidr who was the exclusive patron saint of the festival of Bera Bhashan in the sixteenth century, folklorists note the introduction of other figures upon the stage of the raft by the seventeenth and eighteenth centuries and a shift in meaning of the festival from an invocation of Khidr's protection to appeasing him and other malevolent beings who would do humans harm (Sayidur 1991).

Among the songs that used to be sung during Bera Bhashan was a *panchali* or the ballad of Ganga Devi's desire to marry Khwaja Khijir. This song was played with the child at one's breast, whom the woman loves and desires but who bitterly disappoints. I find a certain resonance here with the figure of the drowned child. In the ballad Khwaja and Madar are two orphans who grow up to become famous pirs or sages. This biography is patterned on the introduction of Islam into Bengal through individual saintly figures and the competition with figures in contending traditions (Eaton 1996). Khwaja acquires sovereignty of the waters. Ganga, who is the sacred manifestation of a river, seethes, presumably at his intrusion into her domain. She demands that Khwaja marry her because she is no longer a maiden. Khwaja demurs, saying that as a fakir or dervish he cannot marry. Ganga takes it upon herself to trick him into marrying her and goes through many guises before deciding on a particular one of herself as a beautiful young woman in a colorful sari. As she goes in search of Khwaja, she finds a child playing in the dust. When she asks after the child's parents, she learns that he is an orphan, or else why would he be found on the ground and not in his mother's lap? She feels such pity for the child that she takes him upon her lap and feeds him milk from her breasts. The child transforms into Khwaja. Ganga is so incensed that she hurls him to the ground, thus killing him, while she sheds tears of shame and regret. However, being a *zinda pir* or immortal, Khwaja revives, and, with him, Ganga's desires are renewed. The ballad ends with the promise that Ganga will marry Khwaja on the Day of Judgment, when everyone will be assembled on the Maidan-e Hashor, the Great Plain—the implication being that the entire Muslim community will witness this union, and every child will be a cherished guest at the wedding. Just as in the little ditty that the women group members would sing as young girls during Bera Bhashan, Ganga Devi lives in anticipation of her groom's imminent arrival. And in one particular interpretation it is said that it is the children who have died early deaths who will be welcomed by Khwaja Khijir and Ganga Devi at their marriage celebrations (Sayidur 1991).

Here I note the first element of paradox that attends to lives in the *chars*: that death by water—the tragic short-circuiting of a young life—is an invitation to a joyous celebration of union. Before I consider the particular knot of this paradox with material life on the *chars*, I want to consider further how these attenuated references to once-vibrant ritual practices opens us to the wider topography of Hindu mythological tales. This allows us to speculate about how Khidr comes to be twinned with Ganga, what Ganga introduces into the mix as both a mythic figure and a river, and how her inclusion deepens the paradox of death by water.

In one of the tales in the Mahabharata, the Sanskrit epic poem written between the years 540 and 300 BCE, Ganga is wedded to King Shantanu, who is disallowed from questioning her actions.[5] As we will later see, this is how Khidr is linked to the Prophet Moses within the Islamic tradition, as one who cannot be questioned, whose actions have to be borne without challenge or one risks losing a guide to divine knowledge. As Ganga births children, she drowns them one by one in the river, an action which bears an inverse relation to the drowning deaths of children in the women's stories in which the children are enticed to the water. Ganga's purpose for killing her children, as she explains to the silent but suffering Shantanu is to spare the children the pain of mortality. Shantanu stops her from killing the eighth child, who grows up to acquire the title of Bhishma or the immortal one and to whose immortal and chaste status Khidr bears a structural resemblance. The influence of this story upon the much later ballad of the joyous union of Khidr and Ganga in Maidan-e Hashor is to further transvalue the ascension to immortal life through death over a life haunted by mortality, rendering the paradoxical twinning of death with joyous celebration, and death with immortality.[6]

This paradox finds its mundane equivalent in the entwinement of the pleasure and the threat of being in water within everyday life. In Sand and Water, a beautiful and meditative documentary on lives on the Jamuna *chars*, the filmmaker Shaheen Dil-Riaz (2004) takes a few moments in the film to focus on young children playing in the water that captures well the paradoxes of tragedy and joy, of pleasure and threat. Among the children playing in the water, one child, perhaps eight years of age, is evidently the caretaker of two other children, perhaps five and two. She takes them ashore, putting them securely on the land before diving back to play with her friends. The two year old is bereft and immediately dives into the water. The five year old jumps in struggling to bring the youngest back. The eight year old must have caught sight of what was happening as one can hear her yelping and thrashing off-screen as she reaches her two wards. She struggles with the two and drags them out of the water, with the two year old holding on with obvious delight.

The threat menacing this image recalls the words of the women group members who told me about Bera Bhashan from their childhood and the waning of this festival in the *chars*. Among them, Sonekha remembered a time when water was not so plentiful as it is today, just a pump of the tube well handle away. This must have been before the early 1970s when tube wells first became commonplace in Bangladesh. It was her *boyu kaal* (life as a young bride) when she had to bring water from far away wells and rivers in order to help her mother-in-law in household chores. When they were short of water and there would be only a cup or so left, she would be tempted to throw some lentils into it to wash the food so as to put to use the last bit of precious water. However, her mother-in-law would stop her, warning that throwing in the lentils was presaging the drowning of her children and would make her throw away the remaining water.

Was not using the remaining water an expiation of the water's vengeful qualities? In actuality, it was the lentils in the water that drew the mother-in-law's horror, as the lentils in the water were a figuration of drowning children, a miniature of helpless bodies in water and of the distracted purposiveness of the mother in putting them there in the first place. In another instance, I heard the women of a household blame a neighbor for the drowning deaths of her three sons because she had not taught them to swim as babies. This condemnation was followed up by saying that she had gone so far as to feed them milk from a feeder rather than from her breasts. These were both taken as signs not only of her overindulgence but also of forgetting her place in the world—of course at the bottom of the social hierarchy for being a *chaura* but, simultaneously, in the middle of the river. In their article, Blum et al. (2009) explain that women were customarily not allowed to attempt to save their drowning children for fear of becoming bewitched. The understanding was that the body was already enchanted and would contaminate them. Perhaps the fear was that the mother would be so contaminated that she would attempt to drown her child, deepening the association of mothering with murderous instincts.

However, Sonekha laughed; nobody believes such *hastor* or sayings anymore. Everybody shrugged and laughed asking who would protect them now from the clutches of Ganga Ma and Khwaja Khijir, suggesting either a skepticism toward the efficacy of such ritual practices or an awareness of their radical vulnerability to their environs and psyches.

Khidr as Varuna/Shiva and/or Khidr as Ganga

While such paradox cannot be too far from the lives of the poor in Bangladesh, the fact that Khidr appears in many instances within the *chars* suggests that his nature is paradoxical and that it is in his nature to proliferate paradoxes. Earlier I mentioned that folklorists consider him to have become the structural equivalent of the Vedic god Varuna in the Bengal context (Haque 1995 [1975]). Varuna is interesting because he has a dual nature.[7] As the god of the waters, he rules through caprice, requiring sacrifices to appease him. At the same time, he rules through law as the lawgiver to the watery underworld. The expiation of Khidr in the ritual of Bera Bhashan speaks to his capricious nature, while another account of him in the *chars* speaks to the fact that he is less lawgiver, since within the Muslim context only God gives law and the prophets bring law to the world, and more the upholder of God's law underwater. As such, he bears Varuna's dual nature. An elderly *chaura* in Dil-Riaz's (2004) documentary film best articulates this second aspect of Khidr's nature:

> When I was a child, I closely observed erosion. Before erosion begins, the water bubbles up like in a rice pot. Churning water pounds against the riverbank. The current becomes very strong. The waves create whirlpools next to the shore. Big chunks of earth crash into the water and are carried away. Our forefathers said that deep in the water lives the Prophet Khwaja Khijir. He has many laborers. They dig under the soil below until big chunks fall down from above. We haven't seen it ourselves, but it is really true. This lord of the water exists and so do his servants. In the winter months they measure how much earth they want to break away in the coming year. And when the monsoon floods come, they fall upon the bank again.

In another instance of this story, I was told that Khidr only does this because God has commanded him. He measures and erodes only as much land as is required to be given in taxes to God—hence his need for links and laborers for careful measuring. These point to the long-standing culture of land measurement for the settlement and resettlement of *chars* as they emerge and erode into the river waters as well as the colonial practice of taxation on land. However, it also points to the encompassment of daily lives by the word of God. Khidr is bound to His word, and through this relationship, we explore the specific nature of the paradox that the Arab Muslim Khidr brings to this watery milieu.

For a less Vedic, later Brahmanic association, Khidr is also linked with the Nath traditions through the founding figure, Matsyendranath, who was said to have been thrown into the ocean by his parents and swallowed by a fish. There he stayed for many years until he heard Shiva's secret advice on the practice of yoga to Parvati and by means of which Matsyendranath became enlightened. Thus, through the Nath tradition, Khidr acquires another duality, that of the creativity and destructiveness of Shiva.

Before we consider the Islamic figuration of Khidr and the paradoxes therein, I want to point to the fact that the structural equivalence between the Vedic deity Varuna and Khidr or between Shiva and Khidr only yields Khidr a dual nature, whereas the involuted relationship between Khidr and Ganga is perhaps generative of further theorizations of the inflexion between Islam and Hinduism in Bengal. This involution produces a Khidr akin to Ganga as the one whose actions cannot be questioned and as one begotten by Ganga in his congruence to Bhishma in the story from the Mahabharata, not only violently weaned at her breast but also sworn to be her consort in the ballad from Bengal. However, rather than assume a structural equivalence between Khidr and Ganga toward his overtaking of her, as Richard Eaton (1996) and Muhammad Haque (1975) might suggest, could one posit a relationship of paradox between them of the kind Carlo Severi (2004) suggests, but a little differently posed, as in: "If I am separate from you, I am in you?" I would further speculate that it is this relationship to Ganga that allows Khidr to have both mythological figurations and to be naturalized as a part of the physical landscape, as being literally within the Ganga in her manifestation as a river. As Steven Darian (1978) writes about Ganga in Bengal and Bangladesh, within the Marufati, Murshidi, and Baul mystical poetry and songs, Ganga is memorialized not as a divine figure but, simply, as a river with a contradictory nature, an archetype for all other rivers. Might then Khidr's separateness from her introduce new strains and tendencies within the river?

Islamic and Greek Prefiguration of Paradox

Although Khidr finds no explicit mention within the Qur'an, his presence has long been sensed in the eighteenth sura or verse, which is called Sura Al-Kahf (The Cave) (Brown 1983; Netton 1992; Omar 1993). The figure of Khidr drawn from this sura fits best that of the spiritual guide within Sufi Islam. In the events recounted in the Qur'anic verses, the Prophet Moses goes in search of the point at which two seas meet. The dried fish that he takes along with him as his meal miraculously regain life and swim away at a location where Moses finds, in God's words in translation, "one of our servants, whom we blessed with mercy, and bestowed upon him from our own knowledge." Moses requests this person, presumably Khidr, to allow Moses to travel with him so that he may learn from Khidr. To this statement, the latter replies: "You cannot stand to be with me. How can you stand that which you cannot comprehend?" Moses promises not to question Khidr's motives.

During the course of his journey alongside this mysterious person, Moses witnesses Khidr carry out inexplicable acts that fly in the face of mercy and wisdom, the very qualities bestowed on Moses by God (Omar 2010). Khidr hacks a hole in the boat of a poor man who had just provided them passage. This seemingly vindictive act is followed by an even more aberrant one of Khidr taking the life of a young boy playing among his mates. Finally, Moses witnesses Khidr restoring the crumbling wall of a house whose inhabitants had just turned the two away without providing them food or refuge.

When Moses can no longer bear the injustice and cruelty of what he is witnessing, he demands an explanation from Khidr. Khidr derides him for his impatience, saying that Moses knows only how to act on the word of the law, betraying his lack of farsightedness and mystical intuition. Nonetheless, he provides Moses the explanations he seeks before dismissing him (Omar 1993). Khidr explains that the boat was best temporarily put out of operation so that it could escape the clutches of an unjust king who was impounding the boats of his subjects. The poor man who had given them passage would not have been able to survive if he had permanently lost his boat. The boy was best killed, as he was going to grow up to perpetrate many evils and bring unhappiness to his parents. While they might grieve him now, they would be rewarded with better offspring. The wall was best repaired so that it might continue to safeguard the treasure hidden within it destined for two young wards of the household when they came of age. If it were found before its time, their greedy guardians would surely have squandered their inheritance.

Although in each of these instances, Khidr has foreknowledge of what is to come and acts accordingly, his actions are inscrutable to Moses. And when Moses is told to accompany Khidr unquestioningly, their relationship, however short lasting, exemplifies for all times that between Sufi initiates and their spiritual guides. Within the history of Sufism, Khidr comes to be one of the few who can initiate people to mystical knowledge by appearing to them in dream or mystical illumination, without the intercession of a living master or the texts of masters (Omar 2010). As a figure close to God, Khidr enables not just direct access to divine knowledge but also the experience of theophany,

a witness to God here robing Khidr in his divine attributes and a paradoxical retort to the problem of theodicy; the question as to why a
benign God allows evil: "In cruelty there is mercy" (Netton 1992).

Beside his figuration as spiritual guide, Khidr appears in other
personas within Islam, in dreams and visions, in solitary form or
intermingled with others, and prophesizing, giving advice, or com-
municating dictates (Longworth Dames 2013). These figurations and
personas also dissolve. For instance, in the Qur'anic story presented
earlier, Khidr comes as a vessel of God's knowledge, offering boons
and dealing death, but by the time this story is retold within the poetry
of Muhammad Iqbal, he is less a vessel and more an aspect of physical
movement, a tendency of time, and a threshold across two conditions,
naturalized as Ganga Devi is naturalized as a river. Iqbal (2003 [1924]),
the "Poet of the East" who was born in the Punjab, the land of five rivers
with which Bangladesh was once affiliated in undivided India and later
in Pakistan, writes of interrogating Khidr in the poem titled "Khidr the
Guide" in Bang-i Dara (Call of the Road), published in 1924:

> To your world-ranging eye is visible the storm Whose fury yet lies in
>   tranquil sleep under the sea
> That innocent life—that poor man's boat—that wall of the orphan
> Taught Moses' wisdom to stand before yours wonderingly.

Khidr's prescience is to be able to see what furies lie beneath calm
surfaces and to alight upon them toward intensifying them. Iqbal's
Khidr's actions as interventions into the eddies of time, as wresting the
truth of life from time:

> *You shun abodes, for desert-roaming, for ways that know.*
> *No day or night, from yesterdays and to-morrows free*

What is the riddle of life? As in the Qur'an, this Khidr also does not
disappoint. He readily replies to this riddle:

> *Constant circulation makes the cup of life more durable*
> *O Ignorant One! This is the very secret of life's immortality.*

Thus, only in movement is life possible. And the perception and appreciation of movement in the midst of apparent stillness, including death,
is Khidr's special attribute (see also Omar 2004). When I ask the *chaura*
men who are often to be found on the banks of the river, sitting and

watching the waters, what they see, they point to the apparent calm surface of the river. They alert me to the *pakh* or eyes that promise whirls of water under the surface, hinting at the treachery of navigating these waters. So it is that many a boat capsizes when it is caught in these eddies, or when it approaches the riverbank, where the current is sometimes most ferocious, creating deeper ditches along the banks than in the middle of the river.

Women are very rarely seen sitting by the river. Instead, they are busy rushing to the water to bathe themselves and their children and washing their clothes and dishes. However, attending to the way they walk, I notice that their feet are almost always flayed, their toes gripping the soil as they walk. So instead of asking them what they see, I ask why they walk this way. They laugh and say that it is because while the land may appear more solid than water, it is more treacherous, constantly slipping away from under their feet. Prosaically, the soil in the *char* does have a sandy quality and requires a fierce grip if one is to move quickly across it. While Khidr and his consort draw our attention to the entwinement of tragedy and joy, the pleasure and threat in water, the dissolve of one in the other, Khidr, in his quality of movement in stillness and of threshold connecting contrasts, directs our attention to the continual motion of sediment and its transition from land to water, and back to land again.

Paradox in the Pathways of the *Char*?

The Muslim Arab Khidr who made an entry into Bengal in the thirteenth century had settled down as a *khwaja/nobi/pir*/dervish with a Hindu goddess as his consort. A self-described lover of Khidr in the *char*, a well-kempt man for these parts, who was dressed in white garb with a bright green turban atop his head of oiled coils of hair, provided me two sermons on Khidr. These suggested how a Qur'anic version of Khidr was being newly taken up in the *chars*, providing another perspective on the women's association of Khidr with Gongima. Khidr's lover had heard these sermons from his spiritual guide, a locally famed pir of Sabri Chistiya affiliation named Maulana Nechari. These are his recollections of the maulana's sermons which describe Khidr in his Qur'anic aspect as guide to the Sufi knowledge of the invisible:

A man goes on a journey with Khidr as he wants Khidr to open his inner eye. Khidr bids him to go to the nearest market to seek out the only man there. He wonders what Khidr means as men belong to markets, and wouldn't they be there in plentiful? But when he gets to the market, he finds that it is crowded with dogs, dogs barking at other dogs, dogs conducting the business of buying and selling and so on. He races around the market until, finally, he sees a man. He approaches the man and inquires after him. The man is grumpy in his answers but says that he is unafraid to give bold answers as he has nothing to fear, for he has divided up his wealth carefully to give one third to charity, one third to his family, keeping one third for himself (this being the usual formula for zakat or charity). So, in meeting this man, the first man realizes that Khidr had indeed given him the inner eye to make the real man visible to him.

A child wants to pursue higher education to become a religious scholar. His family, particularly his father, is cruel to him, forcing him to tend their farm animals. He weeps to his mother who directs him to go sit on the road. There he encounters a person like no other, green in dress with a green turban and flashing green eyes, who asks him what he seeks. When he tells of his aspiration to pursue religious training, the man shows him the road to a Sufi lodge in India where he can get an education. His mother identifies the man in green as Khidr and sends him out into the night to pursue his studies. However, life in the lodge is not easy. He is again bound to a situation of tending cows and goats. When he wearies of ever learning anything, the head teacher bids him to teach his master class. He is too ashamed as he knows nothing, but the teacher draws him to his chest and in so doing transforms him into a scholar. He is henceforth able to teach and so returns home.

These sermons, which bear structural resemblances to the Qur'anic figuration of Khidr as a Sufi guide, may be an indication of the onset of Islamicisation in northern Bangladesh. Yet there has always been an attraction to the reformist zeal within Islam in these parts (Choudhury 2001). In addition, the sermons maybe making visible or rationalizing the long-standing importance of esoteric or *marufati* knowledge that is also intrinsic to these parts (Hatley 2007). This too is nothing new. Curiously, however, although the sermons showcase guidance that comes unexpectedly, through self-revelation, and the working of miracles, paradox is in short order within them. It may be because Khidr is so resolutely male in these expressions, not capricious like Varuna, a conscientious devotee of God undertaking His lawful destruction for Him, or a child on the lap of Ganga. And in these sermons, he is only ever on dry land, secure in its identity as the marketplace or the road to the Sufi lodge in India. If previously it was Khidr who inducted paradox into the *chars*, drawing out the paradoxical nature of living with water, it now appears that it may have been Ganga or the women all along, those living on uncertain grounds, who intensified these qualities within Khidr. And so continues the loop of interpretation between beings and the world, those who draw paradox in, but also make it the preserve of the feminine.

Conclusion

I began by describing puzzling ethnographic encounters with women who had lost their children to drowning deaths. They blamed the deaths of their children on the evil of Gongima. There was no mention of Khidr. The women group members in the *char* spoke of Khidr in the past tense, as one who they used to invoke during Bera Bhashan but not since this festival had lapsed. In only a few places in Bangladesh is it undertaken by specific families, but this too may likely pass. Instead, the fear that was on the rise was that women were to be blamed for the deaths of their children, for wishing, willing, and participating in the deaths. The elderly man in the documentary *Sand and Water* recounted that it was his elders who had told him about this prophet who lives under the water, but he had never seen such a prophet for himself and could give no guarantee of Khidr's existence other than to point to the annual loss of land to erosion.

When I directly ask more women about Khidr, they profess not to know anything, not in the way that they know about Manik Pir, a figure associated with the care of cows. It is worth noting that cows have made a new entry into the lives of the *chauras*, specifically for those women who tend to them in their homesteads. If the women recall a watery figure at all, they speak of Mayicha Dayo, a scaly being who lusts after fish. His name literally translates as "Give Me Fish." He rolls onto fishing boats to lie next to fishermen whispering his entreaties in his ghastly, unearthly, watery voice or he creeps as far as he can onto land to extend the stub of a limb to women to ask them for a share of the fish they are cooking, suggesting a different kind of haunting by lost children. And if there are sites in the landscape that are scary, it is not for their association with Ganga or Gongima but, rather, with Kali, for having once been a Kalighat, a temple of Kali, a goddess more readily associated with death or the consumptive element of life. I was puzzled since Khidr ramifies in so many directions across time and space from the many expressions and gestures, however fleeting, that I had collected from the *char*. I asked the oldest woman I know, one who usually has many fantastical stories to tell about possessions by jinn—God's creations of fire—of men taken to the land of fairies and of Mayicha

Dayo, what has become of Khidr? "He comes from the Book [a reference to the Qur'an] and that is where he lives. We common folk don't have the book knowledge or courage to call on him." So, it would seem that all the while I was sensing his presence in the *char* he was ebbing away. Or it may be that he had ebbed a while ago, but his luminosity was only just fading.

If I am correct in pursuing a Whiteheadian perspective on Ganga and Khidr in the *chars*, this waning suggests something about the women's actual occasions. We have to tread carefully here because we cannot assume a one-to-one correspondence between verbal expressions and empirical reality. The women's actual occasions cannot be reduced to mere spurs to change metaphysical explanations or their narratives as a neutral reflection of their experiences. But if we pursue the element of paradox as a lure from the world, then it would seem that the world does not extend this lure as readily within the women's daily lives as it once did; and that they no longer enliven this lure through their ritual practices, invocations, and remembrances of Ganga and Khidr. And if we previously tracked this lure of paradox to draw out the diverse experiences of watery landscapes and riverine lives and the river's truth for itself, then perhaps water is not as threatening or as perplexing as it once was. I remember the amusement with which my research project was greeted when I first showed up at the *chars*. As I related that I was interested in finding out what it is to live on a river, everyone told me I had missed the river in its fierce persona: "It is now tame. It no longer roars." Perhaps with the ebbing of the enigmatic and the paradox-producing Khidr, we have a hint of the foresight among the *chauras* that the river will be no more? Could its demise also augur the passing of a known world?

Richard Warrick and Qazi Ahmad (1996) are among those who prognosticate the demise of the Jamuna River in Bangladesh. They make the observation that the Jamuna River has already gone from being a perennial river to being a seasonal one. As glaciers melt upriver, there will be a gush of waters through the Jamuna, but this will be for a limited period, and after that there will be no more snow melt to feed its flow. Even as the Jamuna changes its nature, Susan Crate and Mark Nuttal (2009) have shown that environmental change manifesting climatic changes do not simply impact existing forms of life in obvious ways but, rather, are threat multipliers. In other words, climate change ramifies in many directions. I have thus far claimed that the waning of Khidr in the women's expressions tracks the ebbing of the presence of the river in their lives and of the river's own diminution as a physical entity. It is my further claim that climate change also ramifies through the women's modes of thinking and expressing. In a recent article discussing the difficulties of integrating social science with climate change, Francis Moore, Justin Mankin, and Austin Becker (2015) remark on social sciences' discomfort toward the predictive bent of climate science and of its movement between simulation and the real. They resort to the position that climate science must move down in scale to strengthen its generalizations. In so doing, they miss the opportunity to show how social science may attend better to the predictive nature of climate science if it is pointed out that their own interlocutors are given to making such scaled up predictions. While I cannot say if the women's words speak the truth of the eventuality of the demise of the Jamuna from a climate science perspective, I hope to have shown how environmental changes ramified in the direction of women's changing relationship to standing myths, how their mythic and figurative speech provide a sense of the multiplication and even transformation of threats in this milieu, and of how such speech may be given to portending through their action of forgetting. At the same time, given the centrality of paradox to Ganga's and Khidr's mutual modes of being, I wonder if this forgetting is a kind of remembering; if in being forgotten, the river is remembered and if in being remembered, it may yet be re-birthed? Perhaps this is the kind of portending appropriate to the newness of Terra Aqua.

Notes

1 It is important to note that for Alfred Whitehead (1959 [1927]) all experience is emotional, with cognition and consciousness coming later, if it comes at all. Furthermore, he claims that for a proposition or lure from the world to come to constitute an object to the subject or actual occasion of the women, it must be apprehended emotionally, it must come to its fullest attainment as emotion and that this self-unity is the feeling of self-enjoyment or satisfaction. In other words, he is not referring to actual emotions, and in using him, I am not intimating that the women come to enjoy their expressions of the explanation that they give of their children's drowning deaths but that there is an entire process, its emergence, its becoming, its perishing, and its subsequent being for other processes that is enacted by the women's expressions and that make their expressions an actual occasion for them—one looped through the world rather than standing apart from, and commenting on, the world. See also Whitehead's (1967 [1933]) Adventure of Ideas. I thank Steven Shaviro for directing my attention to this source and to Whitehead's Symbolism in support of the arguments I am attempting here.
2 The presumption within Whitehead is that of one world, one of functional activity, in which everything affects everything else with these impingements being the occasions of experiences (see Whitehead 1959 [1927]). Perhaps

this view of the world veers too much in the direction of what William Connolly (n.d.) has called panpsychism, but I think this critique rests upon whether one has a maximal or minimal definition of experience. Having a minimal definition as I do would not obviate the human capacity to have complex experiences but would not make these experiences unique to humans alone, thus releasing them from the burden of being the sole exemplar of higher level consciousness or rather only capable of such. It would allow humans to be expressive of various levels of ambiguity, indistinction, and confusion, making these also constitutive aspects of our experiences. In Symbolism, Whitehead writes about those things that we sense to exert control on us:

> But for all their vagueness, for all their lack of definition, these controlling presences, these sources of power, these things with an inner life, with their own richness of content, these beings, with the destiny of the world hidden in their natures, are what we want to know about.
>
> (Whitehead 1959 [1927], 57)

What I hope to achieve in my explorations of these women's expressions is to show how fragments of a possible shared past with Hindus, now lived in ignorance of that past, in the ruins of once robust myths, also provides the opportunity to see how the women sense controlling presences in their lives and to see, again in Whitehead's words, "descriptions of human experience factors which also enter into the descriptions of less specialized natural occurrences" (Whitehead 1967 [1933], 184). This is how I intercalate the women's initial sense of Gongima's controlling presence with their experience of the river. Claude Levi-Strauss (1992 [1955]) is the anthropologist who comes to mind as having written that myths are the cognitive means by which people try to think and unthink their cultural separateness from nature. While an earlier generation of his readers saw him as someone who subscribes to the separation of culture from nature, even as maintaining that culture creates itself through marking its difference from nature, a more ontologically inflected reading of him by Eduardo Kohn (2015) and Eduardo de Castro (2014) shows him suggesting that thought is immanent in the world. If we follow through the Whiteheadian resonances, this would be akin to Whitehead (1959 [1927], 26) saying: "Abstraction expresses nature's mode of inter- action and is not merely mental. When it abstracts, thought is merely conforming to nature—or rather, it is exhibiting itself as an element of nature."

3 I am thankful to Annu Jallais for making this connection.
4 I am grateful to Swayam Bagaria and Andrew Brandel for drawing my attention to this story and for helping me to think through the relationship between Khidr and Ganga.
5 I note here that this association of death to celebration is most commonly seen in the context of the annual *urs* (literally, wedding indicating the union with God) in the dargahs (literally, courts, indicating shrines) of saints that are densely scattered across the Bengal landscape.

6 The classic statement on Varuna is by Georges Dumezil's (1988) *Mitra-Varuna: An Essay on Two Indo-European Representations of Sovereignty*. Bhrigupati Singh (2015) utilizes this work very effectively to show how state power in India operates through a twinned mode of capricious control and concern for the welfare of people.

7 I am grateful to Projit Bihari Mukharji for this association between Khidr and Matsyendranath. For an introduction to the Nath tradition, see Mallinson (2011).

References

Albinia, Alice. 2010 *Empire of the Indus: The Story of a River*. New York: W.W. Norton and Company.

Amoore, Louise. 2013 *The Politics of Possibility: Risk and Security beyond Probability*. Durham: Duke University Press. http://dx.doi.org/10.1215/9780822377269

Blanchet, Therese. 1984 *Meanings and Rituals of Birth in Rural Bangladesh: Women, Pollution, and Marginality*. Dhaka: University Press.

Blum, Laura S., Rasheda Khan, Adnan A. Hyder, Sabina Shahanaj, Shams El Arifeen, and Abdullah Baqui. 2009 Childhood Drowning in Matlab, Bangladesh: An In-depth Exploration of Community Perceptions and Practices. *Social Science and Medicine* 68(9):1720–1727. http://dx.doi.org/10.1016/j.socscimed.2009.01.020

Brown, Norman O. 1983 The Apocalypse of Islam. *Social Text* 8(8):155–171. http://dx.doi.org/10.2307/466329

Choudhury, Nurul H. 2001 *Peasant Radicalism in Nineteenth Century Bengal: The Faraizi, Indigo and Pabna Movements*. Dhaka: Asiatic Society of Bangladesh.

Connolly, William. n.d. "Distributed Agencies and Bumpy Temporalities" (unpublished).

Crate, Susan A., and Mark Nuttall. 2009 *Anthropology and Climate Change: From Encounters to Actions*. New York: Routledge.

Darian, Steven G. 1978 *The Ganges in Myth and History*. Honolulu: University of Hawaii Press.

De Castro, Eduardo Viveiros. 2014 *The Cannibal Metaphysics*. Translated by P. Skafish. Minneapolis: University of Minnesota Press.

De Pina-Cabral, Joao. 2014a World: An Anthropological Examination (Part 1). *Hau: Journal of Ethnographic Theory* 4(1): 49–73. http://dx.doi.org/10.14318/hau4.1.002

De Pina-Cabral, Joao. 2014b World: An Anthropological Examination (Part 2). *HAU: Journal of Ethnographic Theory* 4(3): 149–184. http://dx.doi.org/10.14318/hau4.3.012

Dil-Riaz, Shaheen, director. 2004 Sand and Water (documentary film).

Douglas, Mary, and Aaron Wildavsky. 1983 *Risk and Culture: An Essay on the Selection of Technological and Environmental Dangers*. Berkeley: University of California Press.

Dumezil, Georges. 1988 *Mitra-Varuna: An Essay on Two Indo-European Representations of Sovereignty*. New York: Zone Books.

Eaton, Richard. 1996 *The Rise of Islam in the Bengal Frontier, 1204–1760*. Berkeley: University of California Press.

Environment and Geographical Information Systems. 2000 *Riverine Chars in Bangladesh: Environmental Dynamics and Management Issues, Support Project for Water Sector Planning*. Dhaka: University Press.

Halewood, Michael. 2014 The Order of Nature and the Creation of Societies. In *The Lure of Whitehead*, edited by Nicholas Gaskill and A. J. Nocek. Minneapolis: University of Minnesota Press: 360–378.

Haque, Muhammad Enamul. 1995 History of Sufism in Bengal. In *Muhammad Enamul Haque Rochonabali*, vol. 4, edited by Mansur Musa [1975]. Dhaka: Bangla Academy.

Hastrup, Kristen, and Cecilie Rubow, eds. 2014 *Living with Environmental Change: Waterworlds*. London: Routledge.

Hatley, Shaman. 2007 Mapping the Esoteric Body in the Islamic Yoga of Bengal. *History of Religions* 46(4): 351–368. http:// dx.doi.org/10.1086/518813

Iqbal, Muhammad. 2003 Bang-i Dara. In *Poems of Iqbal*, translated by Victor Gordon Kiernan [1924]. Lahore: Iqbal Academy: 2–55.

Islam, M. Nazrul. 2010 *Braiding and Channel Morphodynamics of the Brahmaputra-Jamuna River*. Saarbrucken: Lambert Academic Publishing.

Jahan, Rounaq. 1972 *Pakistan: Failure in National Integration*. Dhaka: University Press.

Khan, Naveeda. 2014 The Death of Nature in the Era of Global Warming. In *Wording the World: Veena Das and Her Interlocutors*, edited by Roma Chatterji. New York: Fordham University Press: 288–299. http://dx.doi.org/10.5422/fordham/9780823261857.003.0016

Kohn, Eduardo. 2015 Anthropology of Ontologies. *Annual Review of Anthropology* 44(1): 311–327. http://dx.doi.org/10.1146/ annurev-anthro-102214-014127

Levi-Strauss, Claude. 1992 *Tristes Tropiques* [1955]. Translated by John and Doreen Weightman. New York: Penguin Press.

Longworth Dames, M. 2013 Khwadja Khidr. In *Encyclopedia of Islam*, 2nd edition. Leiden: Brill Online.

Mallinson, James. 2011 Nath-Sampradaya. In *Brill Encyclopedia of Hinduism 3*. Leiden: Brill: 407–428.

Moore, Francis C., Justin Mankin, and Austin Becker. 2015 Challenges in Integrating the Climate and Social Sciences for Studies of Climate Change Impacts and Adaptation. In Jessica Barnes and Michael Dove, eds. *Climate Cultures*. New Haven: Yale University Press: 169–195. http://dx.doi.org/10.12987/yale/ 9780300198812.003.0008

Mukherjee, Rila. 2008 Putting the Rafts Out to Sea: Talking of "Bera Bhashan" in Bengal. *Transforming Cultures eJournal* 3(2): 124–144. http://epress.lib.uts.edu.au/journals/ Tfc

Netton, Ian Richard. 1992 Theophany as Paradox: Ibn al-'Arabī's Account of al-Khadir in his Fusu#s al- Hikam. *Journal of the Muhyid'din Ibn Ara'bi'Soci'ety* 11: 11–22.

Omar, Irfan A. 1993 Khidr in the Islamic Tradition. *Muslim World* 83(3–4): 279–294. http://dx.doi.org/10.1111/j.1478-1913.1993.tb03580.x

Omar, Irfan A. 2004 Khizr-i Rah: The Pre-eminent Guide to Action in Muhammad Iqbal's Thought. *Islamic Studies* 43(1): 39–50.

Omar, Irfan A. 2010 Reflecting Divine Light: Al-Kihdr as an Embodiment of God's Mercy. In *Gotteserlebnis und Gotteslehre: Christliche und Islamiche Mystik im Orient*, edited by Tamcke Martin. Wiesbaden: Harrassowitz: 167–180.

Roy, Asim. 1984 *The Islamic Syncretistic Tradition in Bengal*. Princeton: Princeton University Press. http:// dx.doi.org/10.1515/9781400856701

Sayidur, Mohammad. 1991 *The Festival of Bera Bhasan (Bera Bhasan Utshob)*. Dhaka: Bangla Academy.

Scheper-Hughes, Nancy. 1992 *Death without Weeping: The Violence of Everyday Life in Brazil*. Berkeley: University of California Press.

Severi, Carlo. 2004 Capturing Imagination: A Cognitive Approach to Cultural Complexity. *Journal of the Royal Anthropological Institute* 10(4): 815–838. http:// dx.doi.org/10.1111/j.1467-9655.2004.00213.x

Singh, Bhrigupati. 2015 *Poverty and the Quest for Life: Spiritual and Material Striving in Rural India*. Chicago: University of Chicago Press. http://dx.doi.org/10.7208/chicago/9780226194684.001.0001

Steiner, Franz. 2013 *Taboo* [1956]. London: Routledge.

Stewart, Tony. 2001 In Search of Equivalence: Conceiving Muslim-Hindu Encounters through Translation Theory. *History of Religions* 40(3): 260–287. http://dx.doi.org/10.1086/463635

Strathern, Marilyn. 2006 Divided Origins and the Arithmetic of Ownership. In *Accelerating Possession: Global Futures of Property and Personhood*, edited by Bill Maurer and Gabrielle Schwab, New York: Columbia University Press: 135–173.

Trawick, Margaret. 1992 *Notes on Love in a Tamil Family*. Berkeley: University of California Press.

Warrick, R. A., and Q. K. Ahmad, eds. 1996 *The Implications of Climate and Sea-Level Change for Bangladesh*. Dordrecht: Kluwer Academic Publishers. http://dx.doi.org/10.1007/978-94-009-0241-1

Whitehead, Alfred North. 1959 *Symbolism, Its Meaning and Effect* [1927]. New York: Capricorn Books.

Whitehead, Alfred North. 1967 *Adventure of Ideas* [1933]. New York: Free Press.

Whitehead, Alfred North. 1979 *Process and Reality* [1929]. New York: Free Press.

5 Earth, Water, Salt

Amphibious Pasts of the Lower Gangetic Delta[1]

Sudipta Sen

The Bengal Delta

Tigers of the Sundarbans mudflats do not look like those that live in the mainland forests of India. Some zoologists say that they show distinctive morphological adaptations that equip them for an amphibious, mangrove dwelling (Adam Barlow et al. 2010, 329–331; Mullick 2011). Their heads are smaller, and their stripes less pronounced. Their reduced bodyweight helps them move swiftly over breathing roots and mud, and they paddle across creeks of brackish tidewater, seemingly with little effort. Using tails as rudders, they swim long distances between islands and mudbanks, crossing rivers and channels without trouble. Some have been recorded as swimming for more than 12 miles at a time. While more than one-third of their diet consists of wild boar, cheetal dear, and the occasional macaque, it also includes a substantial amount of crab and fish. Among their prey are also indigent, unfortunate humans who brave the treacherous mangrove belt for honey, wood, and fish. As Annu Jalais has shown in her evocative ethnography of the Sundarbans, not only is the tiger revered as a fearsome spirit of the forest, but people living in its proximity are also believed to possess some of its peculiar attributes (Jalais 2009, 10, 204–206).

Like crocodiles, fish-eating gharials, and monitor lizards, many naturalized fauna of the Sundarbans have become amphibian over time, part of a unique habitat stretching over hundreds of miles—from the fringe of the Gangetic delta in India to the Ganges–Brahmaputra debouchment corridors in present-day Bangladesh. Taken together, the contiguous mouths of the Ganges, Brahmaputra, and Meghna rivers make up one of the largest deltaic formations in the world, the environmental balance of which now is in grave peril. Climate change and global warming have altered the seasonal and diurnal behavior of ocean tides accentuating the submergence of the mudflats. Channels of rivers

DOI: 10.4324/9781003282471-6

change course more capriciously and often. It is estimated that during the eighteenth century the mangrove belt was twice the size of what it is today (Seidensticker and Hai 1983). The depletion of this wilderness along with extensive habitat-loss for species has occurred across two contiguous but different timescales—the long-term advance of human settlement and cultivation, and becoming apparent in recent years, global warming and climate change leading to higher oceanic tides, more devastating cyclones, and toxic levels of salinity. The freshwater flow from the Ganges and its distributaries dwindles every passing year, resulting in greater inflow of tidal bores, disturbing the floral balance of the coast and islands, and shrinking the domain of mangroves. In the last 25 years four islands, Bedford, Lohachara, Kabasgadi, and Suparibhanga have disappeared undersea, and less than two miles from the mouth of the Hugli, much of Ghoramara Island is under water. Thousands of displaced islanders have since joined the ranks of the climate refugees of the Anthropocene.

The amphibian behavior of terrene mammals has for long been a part of the folklore of the Sundarbans. Here, wild boars (*Sus scrofa*)— usually regarded as vegetarian or carrion-feeders—dig up fish that get trapped in the muddy creeks and feed on crabs and other crustaceans. In Narayan Gangopadhyaya's memorable mid-twentieth-century short-story *Dosor* (*Companion*) on the rivalry between a cultivator of *char*s and a wild boar, the veteran, grizzled peasant protagonist marvels about the Sundarbans boar. New sandbanks emerging from the saltwater, where wild vegetation takes root tenaciously despite repeated flooding, infested with poisonous reptiles, are notoriously unfit for terrene fauna, let alone humans. The first mammal to arrive at these newly formed *char*s or banks however, the old man asserts, is always the wild boar. It swims fearlessly and hunts for crabs. When deadly tidal bores sweep in "it stands upright like a human" grabbing on to the branches of the mangrove with its forelegs (Gangopadhyaya, 2010). As the saltwater rises it simply thrusts its snout above the surface.

Today, as the sea rises and cyclones become more frequent and devastating, and salt despoils tillage, destitute humans compete with veteran amphibious rooters, foraging for crabs, shrimp, firewood, and honey, paddling, and wading into the territory of swamp tigers, crocodiles, and bull sharks.

This essay is a brief venture into a deeper history of this land of mudflats, sandbanks, creeks, and channels where hunter-gatherers, boaters, and fishers have populated the margins of fresh and saltwater, river, and ocean, as protagonists of a perpetually glutinous terrain— part earth, part water. It looks at how the lifeworld of such people and

Figure 5.1 Sundarbans wild boar foraging in the mud (Photograph by Debal Sen).

their ambient ecology developed over the *longue durée* along the deltaic edges of Bengal, and how they in turn reshaped their surroundings with hand-crafted vessels, nets, drifts, trawls, bags, fishing spears, along with plows and scythes in the low-lying rice paddies, armed with generational knowledge of ocean tides, monsoons, estuarine biodiversity, and the marine biome. Difficult as it may be, sifting through colonial-era anthropological and census data dating back to the nineteenth century, we can try and reconstruct a vignette of this amphibian terrain and its older human footprints.

At the same time, how do we frame histories of such littoral dwellers that do not fit the description of peasant, nomad, or tribal? How do we account for subaltern entities that the colonial anthropological surveys and censuses routinely rounded up as the Dravidian "boat castes" and the "remnants of a distinct aboriginal tribe" steeped in their marine ancestral myths and rituals dedicated to animist water deities? How do we re-imagine such lives and livelihoods that seem to be inextricably merged with the spatiotemporal singularity of this vast shoreline scored by the ebb and flow oceanic currents, torrential monsoons, and diurnal tidal bores? Their histories, I suggest, can only be articulated through a deep-ecology of the delta as a shape-shifting and volatile geo-historical formation, atop fossil remains of ancient peat-beds, freshwater and marine sediments, and primeval carbonaceous clay. These elements

Figure 5.2 Egret and low-lying mangroves, Sundarbans (Photograph by
          Debal Sen).

of a long-buried past surface in bits and pieces in early colonial land
surveys and geological expeditions, showing that British geographers
and naturalists might have also caught a glimpse of this remarkable
depth of human, faunal and floral habitat and long-term cultural adaptation to jungles, mangroves, riparian channels, marshes, and swamps.

While the future of this coastline is at grave risk today with climate
change and ocean rise, the precarious equations of land and water,
river and ocean, and mud and salt have defined the character of life
and survival here for centuries past. Their evidence shows that adaptive
responses to sea-level rise and burgeoning, turbid storms—outlined
by ethnographers such as Gagné and Rasmussen as the subject of an
amphibious anthropology peculiar to our times—have been in play
in these parts long before the first intimations of the Anthropocene
(Gagné and Rasmussen 2016). The Sundarbans littoral is where the sea
has always returned to claim its share of the land, where freshwater
and silt brought from the plains by the mighty rivers of the subcontinent have run headlong into the murky currents of the Bay of Bengal,
creating alternate rhythms of sedimentation and erosion, helping sustain one of the largest halophytic mangrove coasts of the world. These
strata on which they thrive are neither liquid nor solid. The very logic

of such formations defies any simple or consistent geological pattern (Bremner 2015).

Ancient Peat

There is a plethora of terms for waterlogged terrain in Bengali. The generic word for mudflats is *bādā*, as also in mangrove forest (*bādābana*) or swampy wilderness (*bādāṛa*). *Jalā* derived from *jala* (water) is a marsh, along with *hāor* (var. *hābaṛa*) periodically flooded ground. Among various waterbodies we have the ubiquitous *khāla* or creek, and lakes and fens such as *bila* and *jhīla*. Lowlands of the Bengal delta were of particular interest to the English in India, who had established the city of Calcutta by reclaiming three coastal villages near the mouth of the Hugli River surrounded by malarial swamps (Sinha 2014; Bhattacharyya 2018). James Rennell, who led the early geographical surveys of the East India Company, remarks in his *Memoirs* how rapidly the delta could grow and extend into the sea with the mud and sand transported by rivers from the "remotest ages, down to the present times" (Rennell 1792, 346). He also noticed that because of such activity the sea was turgid almost up to 20 leagues from the coast, and in places the alluvium brought in by the Ganges and the Brahmaputra created sandbanks extending more than 20 miles into the ocean. These rivers continuously formed and swept away islands. A glass of water taken from the Ganges in spate was analyzed by Rennell to be one-quarter full of mud, an indication of the kind of thick deposits receding waters routinely left at the fringes of the Bengal delta.

The vast, muddy wilderness of the Sundarbans never failed to catch the attention of British observers. The renowned naturalist Joseph Dalton Hooker in the 1840s journeying from Noakhali to Chittagong described the narrow creeks and islands of the lower delta as "channels scooped several fathoms deep in the black mud" (Hooker 1891, 535). Traveling for days on end on small boats he kept seeing the "glistening oozy mud" bereft of shrub or tree all the way up to the horizon, broken only by the advancing white line of incoming tidal waves. Such accounts give us some idea of the historic span of these deltaic swamps, and the ubiquity of estuarine mud—a substratum of silt and decomposed organic matter. As British military engineers found for the first time in 1840, the peat content in the alluvium taken from Calcutta (9–25 feet) yielded 62% "volatile" aqueous substance, 21.3% of red ash and 16.7% of fixed carbonaceous matter.[2]

Early British and hydrographers who studied the coastal areas of Sundarbans noted a furiously restless landscape, marked by centuries of volcanic and seismic activity. From the rise of the Rajmahal hills, where the Ganges dips southward toward Bengal, they discovered alternate strata of peaty and vegetal deposits, with remains of trees at great depths. They found alternate elevation and depression of the ground and continual erosion caused by the inroad of tides, cyclones, and oceanic storms. They also found archeological remains of abandoned settlements, where villages and land under cultivation had been lost to erosion and encroachment of the sea. When the military engineers at Fort William, Calcutta were conducting boring operations looking for sources of fresh water near Calcutta in 1836, digging at the depth of 10 feet they found a bed of blue clay, which became darker in color at the depths of 30 to 50 feet (Smith 1846, 6). These turned out to be large sections of peat and decaying fragments of trees, a bulk of which was identified as belonging to different species of *Heritiera* (local Sundri), once the most common mangrove of the Sundarbans—after which the region is named. Examining this stratum of peat and decayed timber, and the extent of the debris of ancient vegetation, the study reported that jungles and low-lying swamps once covered the entire Gangetic delta before the advent of the savannah and regular human habitation. This was further confirmed by the discovery of fossils of lizards and turtles found in carboniferous clay more than ten feet under, lodged in the remnants of an ancient forest cover.

It is difficult to gage what kind of aggregate, long-term impact British colonial landed settlements had on this ancient ecology. The East India Company wanted detailed charts of the channels and islands of the Sundarbans, especially after the permanent settlement of the Bengal land revenue had increased the appetite of the colonial government for new sources of land taxation. The first military sketch surveys of this area were conducted by Captain James Rennell in the 1770s. In 1828, with a view to new settlements, the combined deltas of the Ganges and the Meghna were finally surveyed up to their easternmost limits by William Dampier and Alexander Hodges. Despite strenuous efforts by the Company to extend its revenue base and reclaim these tracts for agrarian settlements, much of the wilderness remained beyond the reach of land prospectors through the first decades of the nineteenth century. While more than 7000 square miles of were declared as the property of the Company state in 1828 reclaimed or brought newly into cultivation between 1832 and 1872, at certain points along the delta, the forest extended least 81 miles inward from the edge of the sea.[3] The land redistributed was mostly covered in jungle, and only about one-seventh of the entire expanse of the Sundarbans had been converted into cultivable tracts by the 1830s. At the same time, in places like Patuakhali and Bakharganj, more than 45% of the forest and swamps had been cleared, and by 1844, 138 surveyed tracts amounting to more than 300,000 square acres in the districts of 24 Parganas and Bakharganj had been reclaimed. Land speculators were hiring tribal laborers, especially woodcutters, from the Chota Nagpur plateau of India or inner parts of Burma to clear vegetation and drain swamps, leading to renewed squatting and illegal encroachment (Sarkar 2010). New settlements followed the access to freshwater, their patterns reflecting the unpredictable flow of channels and changing contours of the islands.

The unfamiliarity and uniqueness of the mangrove delta also reinforced the notion that it lay beyond the pale of normative Indian peasant society. Classified as wasteland from the earliest reports available, forfeited by default in perpetuity by the government, they were to be sold or leased to daring farmers and zamindars under the category of *patitābādī tāluka*s (*patita*— abandoned, and *ābādī*—cultivation).[4] Early nineteenth-century reports assert that all new land in the delta formed by "alluvion and dereliction of waters" since the passage of the first decennial revenue settlement act undertaken by the East India Company in 1790 belonged unequivocally to the ruling government, which also had legal rights to all produce arising from the reclamation of waste. No matter how incremental, the encroachment of cultivation and the rehabilitation of what one popular English author called a "pestiferous salt mangrove-swamp" over the course of the nineteenth century would lead to a colossal loss of the coastal forest canopy (Simson 1886, 115).

The doyen of British–Indian forestry William Schlich would remark in 1875 that the supply of three great sources of timber from the Bengal coast, Goran (*Ceriops roxburghiana* or *Ceriops decandra*), Gangwa, or Genwa (*Excoecaria agallocha*), and most of all Sundri (*Heritiera littoralis* and *Heritiera fomes*) were in severe shortage from overcutting (Schlich 1876, 9–10). Sundri wood was used extensively for beams, carriages, planks, posts, furniture, and firewood, and greatly prized as construction material for boats. These trees were disappearing rapidly from the western part of the Sundarbans, especially in parts close to the imperial capital of Calcutta, wrote Schlich, stressing the urgent need for mangrove conservation efforts to be put in motion. In retrospect, Schich's concerns have turned out to be unerringly prescient. The Sundri, which thrives on firmer and higher ground has all but disappeared from the western Sundarbans. Taken together, this protracted and dismal history of the reclamation, resettlement, cultivation, and deforestation of the Bengal delta begs the question: How did such intrusion affect the future of its original inhabitants— fishers, boaters, hunters, and foragers? This is a difficult question that cannot be answered adequately in this brief essay. However, it provides an essential premise for a discussion of coastal communities and livelihoods during the colonial era.

Mud, Fish, and Tides

Civil servant, statistician, and scholar of rural Bengal, William Wilson Hunter described the Sundarbans as a "vast alluvial plain, where the continual process of land making has not yet ceased" (Hunter, 1881, 468). Despite evidence of sporadic attempts to set up rural settlements dating back to a period before Mughal rule, followed by early British forest reclamation projects pioneered by Collector General Claude Russell and Tilman Henckell, judge and magistrate of Jessore, 1783, later nineteenth statistical surveys of British India show that the northern frontiers of the Sundarbans had barely moved for over 400 years. This sprawl of creeks, morasses, and swamps belonged to people who lived by scouring the profusion of waterbodies that had developed over centuries. They had had the run of this land of mud and saltwater for generations.

Nineteenth-century descriptions of the Sundarbans suggest a perennial traffic of boaters, fishers, and traders, not in the least bound by the sparseness of dry, elevated land. Age-old floating communities had

Figure 5.3 Fishing boat, Sundarbans (Photograph by author).

arisen and flourished along the rivers, creeks, and estuaries of remote sections of the delta. The census of 1872 reveals that at least 14% of the recorded population of the greater Sundarbans comprised of people of fishing and boating castes (Hunter 1875). The Sundarbans soil laden with decayed vegetal matter supported certain species of rice, but they grew only where one could raise embankments and keep out the tow of brackish tidewater. Living near marshes, rural communities had access to large and small varieties of reed, the slender *pāṭi* and the more robust *nala* (*nalakhāgṛā*) out of which various kinds of mats were woven. These were the mats (*pāṭi, chāṭāi, mādura*) on which people slept and sat on all over rural Bengal. The caste names of such people that come up most frequently in the early statistical surveys and censuses of the lower delta are Malo, Bagdi, Kaibarta, Tiyar, Pod, Chandal, along with poor Muslim converts from among the lower castes (Hunter 1868, 37). James Wise, surgeon, and ethnologist of eastern Bengal grouped the fishing castes of Bengal under the generic term Jaliya (colloquial Jele)—a name deriving from *jāla* or a fishing net. While the occupation of fishing was seen generally as a degrading, "unclean" livelihood from the point of view of caste hierarchy, the fishers of Bengal according to Wise were "… remarkable for strength, nerve, and independent bearing", and among them were some of the "finest example of Bengali manhood" in comparison to "the feeble and effeminate inhabitants" one saw in the cities and towns (Wise 1883, 281). For Wise, the fisherfolk of Bengal were the true autochthons of the lower Bengal delta. His description, predictably daubed with contemporary inflections of race and class, is still striking:

> The three fisher castes of Eastern Bengal, the Kaibartta, Malo and Tiyar, are undoubtedly representatives of the prehistoric dwellers in the Gangetic delta. As a rule they are short, and squat, of a dark brown colour, often verging upon black. Although Hindus by creed, they are fond of showy garments, of earrings, and of long hair, which is either allowed to hang down in glossy curls on their shoulders, or fastened in a knot at the back of the head. The whiskers and moustaches are thin and scrubby; the lips often thick and prominent; the nose short with the nostrils expanded. The physiognomy indicates good temper, sensuality, and melancholy rather than intelligence and shrewdness.

(282)

H. H. Risley, a pioneer of the British anthropological survey in India, supported Wise's findings in his official ethnography of Bengal and argued that the fishing and boating castes of Bengal, such as the

Malo, were not just occupational groups but "the remnant of a distinct aboriginal tribe" (Risley 1892, 65). There is some relevance to these observations, as various Brahminical texts, especially those dating back to the Pala and Sena dynastic periods in Bengal (ca. eighth–twelfth centuries) describe the generic Kaibarta fishers and boaters of the Bengal delta and assign them the lowly status of "corrupted" Shudras or *asataśūdra*s (Ray [1949] 2007). Kaibartas, one of the oldest recorded lower-caste groups of deltaic Bengal, rose in rebellion against the tyrannical Pala-dynasty ruler Mahipala II during the eleventh century (Majumdar 1971). Mahipala was killed and the Kaibarta chiefs Divya, Rudak, and Bhim took over the region of Varendra for a short while before the Palas regained power. This incident alone shows the kind of armed intransigence and political influence the indigenous fishers and boaters might have wielded along the Gangetic basin in Bengal.

Long-term livelihoods tied to fish in the old deltaic settlements of Bengal had contributed to distinct markers of socio-occupational status. The Malo fishers, it is said, branched off from the wider Kaibarta caste group, as did Mech Chandals from the greater Chandal caste who were the traditional disposers of the dead (Hunter 1875). Similar instances can be found among the Bagdis, some of whom were palanquin-bearers and cultivators, alongside Pods and *Chunaris*—lime-manufacturers, the prefix *mechuā* or *mecho* (fishing) being added to their caste name. Some were distinguished by the kind of fishing nets that they made and used. Kaibartas, Tiyars, and Malos wove old nets out of hemp steeped in the glutinous sap of the Gab tree (*Diospyros malabarica*), pounded, and left to ferment and dry in the sun until it toughened and took on a dark color—at least until the mid-twentieth century when they began to be replaced by synthetic nylon (Wise 1883). In his unforgettable 1957 novel *Ganga*, Samaresh Basu wrote about the mythical ancestor of the Malo fishers of the Sundarbans delta who emerged from the ocean depths, with jet-black, curly hair and dark skin glistening with brine, pronged fishing spear in hand. The novel lays out the story of an aging fisherman at the end of his life. Through his eyes we see the treacherous mangroves and the unforgiving sea, and the difficult, fragile lives of his kinfolk who live at the mercy of the spirits of water—the ocean, the sea, and the clouds, keeping close track of the spawning and migratory patterns of estuarine fish such as Bata (*Labeo Bata*), and especially Hilsa (*Tenualosa ilisha*) abounding in the saltwater close to the edges of the Gangetic delta (Hamilton 1822).

Bengali fishers, James Wise thought, were not only "familiar with the habits of fish" but had an instinctive knowledge of the entire deltaic terrain, and much could be gleaned from them about the natural history of the region (Wise 1883, 284). The official count of people who earned a living boating and fishing in Bengal in the 1880s was over 1.3 million (Wise 1883, 285). Wise considered this an unacceptably low number considering how central fish was to the diet of people living in the province. Novelist Basu, who set out with Malo fishers of the Sundarbans to observe their trade closely, wrote poignantly about their struggle with disease and poverty, and their hopes and fear tied to the fish that they caught and killed and owed their lives to, epitomized by the fish-God with unblinking eyes (*khokāthākura*) who watched over them. Native-born to the low-lying and salty channels of the Ganges delta, where they had plied for generations with their narrow, agile crafts and fishing tools, Malos spent their lives in and out of water. Handling nets of hemp and nylon, as their fingers and palms cracked, bled, and stung from the salt, they applied the gum of the Gab tree to caulk their broken skin. For Basu, they were as much a part of the Terra Aqua as the mangroves themselves.

Salt and Sweat

A bulk of the salt manufactured in Bengal came from the salt pans of the lower delta and the mangrove belt.[5] The East India Company had established a strict monopoly on the manufacture and sale of salt. Some of the most lucrative contracts were handed to British speculators and their Indian agents. The price of salt in the early nineteenth century in India was roughly four times than that in London. Salt was made in shallow troughs by the sea along a vast stretch of coastal Bay of Bengal, from Chittagong all the way down to Jaleshwar in Balasore, Odisha. The people who labored in the salt pans were known by their occupation as Malangis who ranked among the poorest people in Bengal, especially in the Sundarbans, where many of them were tethered to their labor through draconian caste rules (especially the *ajūra malāṅgi*) and condemned to hereditary servitude (Mukherjee and Dasgupta 1954, 15). Amounting to about 125,000 souls in southern Bengal, they labored under grueling conditions, subject to various forms of coercion and exaction by salt agents and local zamindars who shared in the profits from the salt trade and the leasing of saltpans (*khālāṛi*s), which were under colonial monopoly (Barui 1985).

Such was the wretched state of the Bengal "Molunghees" and their working conditions that stories of their suffering under the Company's salt monopoly caught the attention of the British parliament in 1796 during the debates conducted on the East India budget.[6] Their plight became the subject of intense scrutiny once again during the public arguments over the renewal of the Company's charter in 1830. John Crawfurd, a civilian who had served in India, Penang, and Singapore thought that they lived in a "virtual state of slavery" (Crawfurd 1830, 9). Others described them as bondsmen forced to make salt in exchange of meager advances for generations, defrauded, forced to work, exposed to the elements—brine, heat, and humidity, and hunted by tigers and crocodiles. Many who tried to escape were caught and brought back to their state of misery by salt agents. The noted Bengali reformer Rammohun Roy sent on deputation as an emissary of the Mughal Emperor to London in 1832 was asked about the salt monopoly— should it stand or should the market in salt be opened to all British merchants. Was their condition better or worse than the poorest of peasants? His answers did not please the Company stalwarts. The Sundarbans salt-makers worked in inhumane conditions battling heat and damp in the mangrove, Roy testified, with little promise of health or "personal freedom", exploited not only by British officers of the salt agencies, but Indian agents as well, and not even spared by the heads of their own caste groups.[7] Monopolists sought to rebut Roy's allegations by soliciting new reports from members of the Boars of Customs.[8] It was not only the makers of salt who suffered in the remote coastal areas of the Sundarbans they argued, but woodcutters, wax gatherers, hunters, and fishers who were also routinely exposed to wild beasts such as tigers and crocodiles.

Why were the Malangis, as contemporary reports asked, "compelled to labor in the pestilential marshes formed by the estuary of the Ganges?"[9] The coastal areas were considered by the British to be the worst parts of the unhealthy province of Bengal. The problem lay with the terrain itself where salt was made. If the forest could be cut down, marshes drained, and replaced with rice—just as had been done in the Sagar Island—the coast could be made salubrious and habitable. Bengal civil servant Thomas Bracken testifying in front of the Select Committee at the House of Commons spoke candidly about the unhealthy conditions in which salt was manufactured in the Sundarbans.[10] Perhaps less frequently than before, but the Malangis were still being carried off by tigers. The committee wanted to establish a "comparative view of death" in the unhealthy districts along the coast, where people had such diseased and famished appearances, and so little to live on.[11] It was not any one policy of the Company, defenders of the salt monopoly argued, but the peculiar nature of the vegetation and swampy terrain that took such deadly toll on the lives and livelihoods of natives.

"Can you compare the waste of human life in the manufacture of salt with the waste of human life in the cultivation of rice, or the manufacture of indigo?"—Bracken was asked by the Company elders.[12] The salt pits closest to the jungle happened to be the most lucrative, especially along the edges of brackish creeks and nullahs that ran through the mangroves. Bracken provided vivid descriptions of the "swampy character of the country" that produced agues and fevers of the "most virulent description".[13] Large quantities of vegetal matter rotted in the water in which the Malangis stood and worked, exposed to the intense tropical sun. One day, he hoped, the Sundarbans would be cleared of jungles and swamps and made fit for proper human habitation.

Enduring Lifeways

The amphibious denizens of the Sundarbans in colonial accounts appear almost indistinguishable from the coastal landscape, interspersed among the mangroves, rivulets, and creeks, their survival epitomized by the diseased, emaciated, and expendable bodies of the Malangis. The salt that they produced, and on which returns of the monopolists depended, helps crystallize both the briny substrate of this amphibious world and the extractive efficiency of the labor regimes and profit margins of colonial venture capitalism. Thomas Wise, a pioneer of colonial medicine and an early observer of race and caste writing around the mid-nineteenth century found that the "hot and enervating" climate of Bengal bred a peculiar kind of people of low stature and agile movements "capable of enduring great fatigue" (Wise 1860, 113).[14] Although endowed with little courage or faculty of mind, they were cunning and surprisingly resilient. "Much exposed to labour" and the cruelties of weather, despite scarcity of food and poor diet, they were somehow able to survive in this terrain of enervating malaise (Wise 1860, 91–92). Read differently, Wise's words can be read as a begrudging tribute to the industry and ingenuity of the people of lower deltaic Bengal, whose struggles long antedated European maritime and colonial presence. One could argue that such resourcefulness and endurance may no longer be adequate today as the sea and cyclonic storms encroach furiously into these spaces, drowning islands and coasts, pouring new salt and marine effluvia into villages and townships—warnings of more severe climatological and demographic upheavals to come.

Fathoming the overlying epochs of history at the edges of the greater Bengal delta demands a new acknowledgment of the incompatible narratives of the late Holocene and the Anthropocene. Anna Tsing has written evocatively about the replenishment of Holocene ecologies as a constant that frames the very idea of human continuity on the planet (Tsing 2017). The amphibian versatility of the inhabitants of the Sundarbans that have endured both the caprice of natural forces and the unaccounted ravages of the British–Indian colonial economy through ingenious mimicries of the aquatic terrain and the sanctuary of the inhospitable mangrove canopy cannot be extended much further in this looming hour of climate crisis, with new forms of ecological breakdown, upheavals, dislodgment, and unprecedented volumes of human and nonhuman refugia. I emphasize *endurance* here, which includes the constants of hunger and disease, rather than sustenance—especially as the acceleration of climate change poses a vital question to any history of survival along the delta. In what ways can we measure the legacy of colonial ecologies as coextensive with ravages of the Anthropocene, especially the extent to which the aqueous and muddy kinship of people, plants, and animals of the Sundarbans was able to stave off the taxation, tillage, and cultivars of the extractive British Indian agrarian machine of the upland watershed of the Gangetic valley? As vital inter- and intraspecies kinships—along with what Donna Haraway has aptly described as the "assemblage of organic and abiotic actors"— inevitably fall apart, so do the fierce and protective barrier of elements and the earthen, vegetal, and aqueous substrates that have sheltered the hominine protagonists of the Terra Aqua for such a long time in this corner of our planet (Haraway 2015, 100).

Epilogue

During a recent visit to the Sundarbans in search of fish-eating wild boars and fishers that brave forest guards and swamp tigers to make a daily living, I was reminded of the resilience and longevity of the mangrove habitat where the semi-lucent river currents and glutinous estuarine silt meets the rise of oceanic tides. Emily Eden, novelist and sister of George Eden, the governor general of British India, traveled to the Sundarbans in the autumn of 1837 and described a very similar landscape of muddy creeks and stunted trees, infested by tigers and poisonous reptiles (Eden 1867). She had never witnessed such a "desolate scene" in her life. Even the sight of a solitary fishing boat struck her somehow as a residue from an antediluvian past—an extension of the timeless expanse of the muddy and brackish tideland. For Eden, the Sundarbans were the remnant of a primeval world "left unfinished when land and sea were originally parted" (Eden 1867, 3). Every now and then she saw a bamboo post adorned with a tuft of leaves, marking the spot where a tiger had made a human kill. It is remarkable how this imagined human absence has pervaded colonial and postcolonial accounts of the Sundarbans. The mangrove forest and swamplands Eden came across during less than a day's boat ride from Calcutta have disappeared today, taken over by the outskirts of metropolitan Kolkata, smaller townships, rice paddies, fish farms, and hatcheries. And yet, the image of the mangrove coast as a perennially inundated wasteland—home of the endangered tiger, which Eden saw almost two centuries ago, persists.

Today, watching the incoming tide lap up the sides of the creeks of the Sundarbans interior—in places like Pirkhali, Harinbhanga, Surjomoni—swirling around the pneumatophores and prop roots, seeping into old tiger pugmarks and gouges left in the mud by foraging boars, one is reminded of the palimpsestic overlap of ecology and survival, of ante-human, diluvial, planetary time and the time of hunter-foragers, fishers, honey-gatherers, and woodcutters. One is also confronted then with the truth of time running out for the people of the Terra Aqua as the speed of sea level rise outpaces their hard-earned habits of survival.

Notes

1 I would like to thank my friends May Joseph and Rohan D'Souza for their ideas and support during the early stages of the writing of this essay.
2 These figures are taken from the "List of Various Boring Experiments Made at Calcutta from 1804 to 1840," *Calcutta Journal of Natural History*, vol. 1, 1841, pp. 327, 340.
3 See the notes of A. D. B. Gomess, Commissioner of the Sundarbans, "On the Progress of Cultivation in the Sundarbans," in H. Beverley, *Report of the Census of Bengal 1872*, Calcutta: Beverley, 1872, p. xii.
4 For a discussion of these transactions see Sarada Charan Mitra, *The Land-Law of Bengal*, Calcutta: Thacker, Spink, 1898, p. 53. *Patitābādī* became a standard term for leases of land under forest cover (*jaṅgalabāṛī*) offered free for a stipulated number of years and then subject to progressive rates of taxation. The legal implications of such deeds are detailed in H. H. Wilson, *A Dictionary of Law Terms*, Madras: Higginbotham, 1891, p. 323.
5 Official estimates from the 1833 suggest that at least one-tenth of Bengal salt came from the salt-makers of the Sundarbans. See *Appendix to the Report from the Select Committee of the House of Commons on the Affairs of the East-India Company, 16th August 1832, and Minutes of Evidence*, London: J. L. Cox, 1833, p. 978.
6 See the "Debate on the East India Budget" [1796] in William Cobbett, *Cobbett's Parliamentary History of England*, London: R. Bagshaw, 1806, p. 1399.
7 Some of Roy's testimony is included the *Appendix to the Report from the Select Committee* [1832], pp. 956-957.

8 Ibid., pp. 976–978. See also G. Chester and H. Sargent's report to the Board
   of Customs, Bengal, 26 Jan 1832.
9 Ibid. p. 978.
10 See Thomas Bracken's testimony of 25 July 1832 in the *Minutes of Evidence
   Taken Before the Select Committee on the Affairs of the East India Company*,
   London: House of Commons, 1832, pp. 270–271.
11 Ibid., p. 270.
12 Ibid.
13 Ibid.
14 Thomas Alexander Wise, *Commentary on the Hindu System of Medicine*,
   London: Trübner, 1860, p. 113.

References

Baird Smith, Richard. 1846 On the Structure of the Delta of the Ganges,
   Exhibited by the Boring Operations in Fort William, A. D. 1836-40. In the
   Proceedings of the Geological Society of London, vol. 4. London: R. and J. E.
   Taylor.
Barlow, Adam C. D. et al. 2010 A Preliminary Investigation of Sundarbans
   Tiger Morphology. *Mammalia* 74: 329–331.
Barui, Balai. 1985 *The Salt Industry of Bengal, 1757-1800: A Study in the
   Interaction of British Monopoly Control and Indigenous Enterprise.* Calcutta:
   K. P. Bagchi.
Bhattacharyya, Debjani. 2018 *Empire and Ecology in the Bengal Delta: The
   Making of Calcutta.* Cambridge: Cambridge University Press.
Bremner, Lindsay. 2015 Muddy Logics. In *Bracket 3: At Extremes*, edited by
   Lola Sheppard and Maya Przybylski. Barcelona: Actar: 199–206.
Crawfurd, John. 1830 *An Inquiry into Some of the Principal Monopolies of the
   East India Company.* London: J. Ridgway.
Eden, Emily. 1867 *'Up the Country': Letters Written to Her Sister from the
   Upper Provinces of India.* London: Richard Bentley.
Gagné, Karine and Mattias Borg Rasmussen. 2016 Introduction—An
   Amphibious Anthropology: The Production of Place at the Confluence of
   Land and Water. *Anthropologica* 58(2): 135–149.
Hamilton, Francis. 1822 *An Account of the Fishes Found in the River Ganges and
   Its Branches.* Edinburgh: Archibald Constable.
Haraway, Donna. 2015 Anthropocene, Capitalocene, Plantationocene,
   Chthulucene: Making Kin. *Environmental Humanities* 6(1): 159–165.
Hooker, Joseph Dalton. 1891 *Himalayan Journals: Or, Notes of a Naturalist in
   Bengal, the Sikkim and Nepal Himalayas, the Khasia Mountains.* London:
   Ward.
Hunter, William Wilson. 1868 *The Annals of Rural Bengal.* London: Smith, Elder.
Hunter, William Wilson. 1875 *Statistical Account of Bengal,* vol. 1. London:
   Trübner.
Hunter, William Wilson. 1881 *The Imperial Gazetteer of India,* vol. 8. London:
   Trübner.

Jalais, Annu. 2009 *Forest of Tigers: People, Politics and Environment in the Sundarbans*. New Delhi: Taylor and Francis.

Majumdar, Ramesh Chandra. 1971 *History of Ancient Bengal*. Calcutta: J. Bharadwaj.

Mukherjee, Tarit Kumar and Arun Kumar Dasgupta, eds. 1954 *Midnapore Salt Papers: Hijli and Tamluk, 1781-1807*. Calcutta: West Bengal Regional Records Survey Committee.

Mullick, Jayanta K. 2011 Status of the Mammal Fauna in Sundarban Tiger Reserve, West Bengal–India. *Taprobanica*, 3(2): 52–68.

Narayan, Gangopadhyaya. 2010 *Galpasamagra: Akhanda Samskarana*. Calcutta: Mitra and Ghosh.

Ray, Niharranjan. 2007 [1949] *Bangalira Itihasa: Adi Parba* [6th ed.]. Calcutta: Dey's Publishing.

Rennell, James. 1792 *Memoir of a Map of Hindoostan*. London: W. Bulmer and Co.

Risley, Herbert Hope. 1892 *The Tribes and Castes of Bengal: Ethnographic Glossary*, vol. 2. Calcutta: Bengal Secretariat Press.Sarkar, Sutapa Chatterjee. 2010 *The Sundarbans: Folk Deities, Monsters, and Mortals*. New Delhi: Social Science Press.

Schlich, William. 1876 Remarks on the Sundarbans. *The Indian Forester*, 1: 6–11.

Seidensticker, John and Md. Abdul Hai 1983 *The Sundarbans Wildlife Management Plan: Conservation in the Bangladesh Coastal Zone*. Gland: International Union for Conservation of Nature and Natural Resources.

Simson, Frank B. 1886 *Letters on Sport in Eastern Bengal*. London: R. H. Porter.

Sinha, Nitin. 2014 Fluvial Landscape and the State: Property and the Gangetic Diaras in Colonial India, 1790s-1890s. *Environment and History* 20(2): 209–237.

Tsing, Anna Lowenhaupt. 2017 A Threat to Holocene Resurgence Is a Threat to Livability. In *The Anthropology of Sustainability: Beyond Development and Progress*, edited by Marc Brightman and Jerome Lewis. New York: Palgrave Macmillan.Wise, James. 1883 *Notes on the Races, Castes, and Trades of Eastern Bengal*. London: Harrison and Sons.

Wise, Thomas Alexander. 1860 *Commentary on the Hindu System of Medicine*. London: Trübner.

Index

For Product Safety Concerns and Information please contact our EU
representative GPSR@taylorandfrancis.com Taylor & Francis Verlag GmbH,
Kaufingerstraße 24, 80331 München, Germany

Printed and bound by CPI Group (UK) Ltd, Croydon, CR0 4YY
11/04/2025
01844012-0004